The Patrick Moore Practical Astronomy Series

More information about this series at http://www.springer.com/series/3192

Building
a Roll-Off Roof
or Dome
Observatory

A Complete Guide for
Design and Construction

Second Edition

John Stephen Hicks

 Springer

John Stephen Hicks
Keswick, ON, Canada

ISSN 1431-9756   ISSN 2197-6562   (electronic)
The Patrick Moore Practical Astronomy Series
ISBN 978-1-4939-3010-4   ISBN 978-1-4939-3011-1   (eBook)
DOI 10.1007/978-1-4939-3011-1

Library of Congress Control Number: 2015947505

Springer New York Heidelberg Dordrecht London
© Springer-Verlag New York 2016

Cover illustration: Cover Photo by Gordon Rife, Astrophotography, Ontario, Canada

Printed on acid-free paper

Springer Science+Business Media LLC New York is part of Springer Science+Business Media (www.springer.com)

This second edition, including dome observatories, is dedicated to my longtime friend and mentor in astronomy, Jack Newton. Jack is to be credited with introducing me to the fascinating study of astronomy and the long journey over 35 years I have had in discovering the wonders of the night and daytime sky. The study of astronomy has opened windows to new friends, star parties, and eclipse travels that have enriched my life. This book is the culmination of 35 years of progressive design of observatories, by trial and error, to what I believe are the two most popular designs, most sought-for by amateur astronomers.

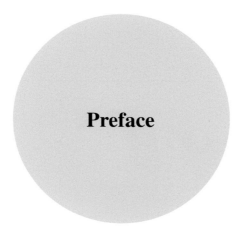

Preface

Building a Roll-Off Roof or Dome Observatory has required several years in all to write, redrafting many diagrams and descriptions to arrive at the most practical and universal model illustrated in this book. At various stages, it was delayed for want of more information on existing observatories and at others carried forward again by a rush of newfound techniques. The prevailing private observatory models you see today involve the roll-off roof mechanism, but the ultimate goal in observatory ownership is the dome observatory—the historically recognized structure. There are a multitude of designs invented to achieve the same result, but only a few good ones, which consumed most of my research effort. Also, all things considered, no one writes a technical book on his own, and at various stages I happily reflect on the people who inspired me to undertake this work.

My beginnings in astronomy were unquestionably launched by the Royal Astronomical Society, Toronto Centre, almost 20 years ago, and by a single individual, Jack Newton, whose extraordinary enthusiasm for astrophotography and observatory building swept me up into a new pursuit. As years went on, attending the annual—Starfest—astronomy convention hosted by the North York Astronomical Association, and the Huronia Star Party Convention in Ontario, Canada, I was an occasional speaker talking about the design of observatories. Various refinements in the construction of these prototypes led to the ultimate design and creation of my own models that have stood the test of time. My wife, conscripted to help in construction, was indeed patient and understanding to put up with the countless hours of diagrams produced at the dinner table, and the geometry that was required to design such structures. As the observatory dome took shape from a skeleton of curved aluminum angle, she was always there to hold a wrench or brace a rib while it was riveted to the curved panels. Many a wrench or rivet gun found its way into the forest, hurled in frustration well into its interior from the lofty dome. It has always been a mystery to me how she endured those hot humid worksessions with the sun's reflection so strong it burnt our eyelids. She rightly deserves my fullest respect for her devotion to a task that was essentially only in "my mind's eye." My first observatory was a domed observatory, entirely self-designed. We named it "New Forest Observatory" because of its location in the center of our magnificent pine forest (Figs. P.1 and P.2).

The "prototype" turned out so well that in fact it became an extension of the telescope you might say. Inside, on sunny mornings, searching the surface of the Sun with my Lunt and Coronado hydrogen filters, I felt quite removed from the clamor of the rest of the world and found real adventure surrounded by the "machinery" of the observatory. Lured by its presence against the forest backdrop,

Fig. P.1 John Hicks beside his observatory within a 20 acre pine forest. (Photo by John Hicks)

I found it hard to do my usual work, and many a client took second place to an observing run in my beautiful observatory. As time went on with improvements, I felt compelled to share the design with others, so I embarked on selling plans for the structure far and wide. Little did I know that hundreds of plan sets would find their way around the world to places like Iceland, Africa, India, and even Australia. It was a real pleasure to share my inspiration with others just starting on the same journey. My first design for a real customer was a request from Don Trombino in Deltona, Florida, who asked me to design a Roll-Off Roof Observatory for solar astronomy.

Don and I were both avid solar astronomers with special requirements for refractors, especially long-focus refractors, which suited the incredible hydrogen alpha filters produced by Daystar Corporation, then in California. The model in this book closely mirrors that which I designed for

Fig. P.2 John Hicks inside the dome beside the main instrument—a 102 mm refractor. (Photo by John Hicks)

Don, which he dedicated as the "Davis Memorial Observatory" in memory of a close friend. We wrote articles together on "Shooting The Sun" and published articles and photos over the years in *Astronomy* Magazine (Figs. P.3 and P.4).

Don Trombino passed away in 1995, and I must believe his observatory still stands even as an elaborate garden pergola. I say this because he had self-designed an attractive patio under the gantry which held the rolled-off roof. He entertained many prominent guests under this enclosure celebrating his new observatory. Among them was Sir Patrick Moore. The most satisfying aspect of promoting a design is the conversations with clients who seek your help. Many situations arise that one never expects. Take the case of a Manhattan astronomer who had cut a hole in his roof for the observatory.

He called me in desperation with the building inspector and fire marshal at his door. They were demanding to know just what he was up to, perforating a large part of his roof. In a hysterical mood, he passed the telephone over to the fire marshal who inquired of me just what kind of structure the man was adding to his roof. I replied "an observatory of course!" I explained that I was a designer and supplier of plan kits for observatory structures all over the world. The "all over the world part" seemed to appease him somewhat, but he still wanted to know what qualifications I had. He was of course concerned over the draught potential created by the open roof in case of a fire. But I quickly informed him that it had to be sealed with a floor and trap door, in order for it to function properly anyway, as it must remain at ambient air temperature. This precaution seemed to provide him with the confidence he needed, and he left the telephone.

I presume all went well from there on as I was never asked for a set of plans by the fire marshal, or any other Manhattan agency.

I cannot overlook the inspiration which followed, from my good friend Walter Wrightman, endowed with a talent in both craftsmanship and inventiveness. Walter walked into my life after I had completed my first domed observatory, wanting to build one of his own. He was well on into old age, suffered from diabetes, and some disability in walking. Yet bounded by these restrictions he designed and built, by himself, a most unique domed observatory. Walter had no formal education past grade 8, drove a cement truck most of his later working life, and took up the science of astronomy like no one I have ever met. Walter and I spent luncheons and coffee breaks for the next few years discussing ways

Fig. P.3 Sir Patrick Moore visits Don Trombino's Observatory in Deltona, Florida. (Courtesy of Jeff Pettitt)

and means of creating better observatories. We went to star parties far and wide to glean ideas, studied the night sky together, and were both so proud of the two observatories we had created. We became known locally, at least around the Town of Newmarket, Ontario, as the Observatory Specialists. We even talked about prefabricating the domes worldwide and traveling to exotic places to site them, all the while meeting the enthusiastic people caught up in a similar rapture. It never happened. Walter eventually died almost blind and unable to move outside the room that imprisoned him. He never stopped talking about observatories nor the various ways to improve the structures. Eventually, his observatory was sold to a friend, who was happy to buy such an exceptional model. I miss his friendship, and his overwhelming devotion to astronomy and the building of observatories.

Fig. P.4 Don seated in the observatory finishing a solar observation and reading his notes. (Courtesy of Jeff Pettitt)

I cannot help but credit him with being the second most influential person toward the writing of this book. I hope such inspiration spreads over to you, the prospective—inventor—of such a structure, for inventor you will be, certainly in the eyes of others who may watch you build it. And remember, before you dedicate its use to just an observatory, it will also serve as a great garden patio enclosure, thanks to my innovative friend, Don Trombino. I credit Don also for most of my inspiration for writing and the idea of launching a book. My only regret is that he is not alive to finally see it. He would have been really proud to see his observatory that he treasured so well, finally in print.

Keswick, ON, Canada   John Stephen Hicks

Acknowledgments

Photo Credits

Throughout the United States and Canada, I searched for owners of Roll-Off and Dome Observatories who had applied their skills to create designs that I would recommend to readers. In the Second Edition that search expanded to retail suppliers of Roll-Off Roof and Dome Observatories to provide options for those astronomers who felt reluctant to build their own observatory, or at least all of it. That search consumed most of my time in the process of writing this book, and proved to be a necessity. A design/build book would be uninteresting without images of actual observatories owned mostly by amateurs—people who are not by profession carpenters, builders, or contractors. Their innovative ideas sparked a lot of my enthusiasm for the task that lay ahead of me. Frankly, their creativity opened paths for many new ideas on techniques that I will use in future designs. For their generosity in supplying photos and descriptions I am most thankful. I know they will feel gratified seeing their particular model exhibited in the book, inspiring others with new construction ideas.

The following observatory owners and suppliers offered photos and/or assistance:

Jack Newton (Arizona Sky Village), Arizona
Donald Trombino (1998), Florida, USA
Richard Kelsch, Ontario, Canada
Gordon Rife (Cover Photo) Ontario, Canada
Mike Hood, Georgia, USA
Bob Luffel, IDAHO, USA
Jay Ballauer, Texas, USA
Greg Mort, Maryland, USA
Larry McHenry, Pennsylvania, USA
Rob Bower, Keswick, Ontario
Gerald Dyck, Massachusetts, USA
Jeff Pettitt, Florida, USA
Terry Ussher, Ontario, Canada
Guy Boily, Quebec, Canada

Doug Clapp, Ontario, Canada
Dave Petherick, Ontario, Canada
Paul Smith (1995), Ontario, Canada
Danny Driscoll, Ontario, Canada
Andreas Gada, North York Astronomical Association (NYAA), Ontario, Canada
Calvin Cassel Jr. USA
Alan Otterson (Albuquerque Astronomy Society, USA)
Walter Wrightman, Ontario, Canada
Brian Colville, Maple Ridge Observatory, Ontario, Canada
Art Whipple, Maryland, USA
Tatsuro Matsumoto, Japan
Jerry Smith, Home Dome Observatories, USA
Gary Walker, Pulsar Observatories, Norfolk, UK, and USA
Wayne Parker, Sky shed Observatories

Products

Companies and Suppliers that offered brochures and advice on their products include:

Andex Metal Products, Suppliers of Sheet Steel Roofing, Exeter, Ontario
Layton Roofing, Keswick, Ontario, Canada
Bestway Casters, Gormley, Ontario, Canada
Harken (Sailing Blocks, etc.) Wisconsin, USA
Schell-Ace Building Centre, Sutton, Ontario
Sky Shed Observatories, Ontario, Canada
Pulsar Observatories, Norfolk, UK, and USA
Home Dome Observatories, USA
Astro Engineering USA
Skywatcher Telescopes Ontario, Canada
Losmandy Astronomical Products, USA
DeMuth Steel Products (Domes), Ontario, Canada and USA
Arizona Sky Village, Portal, USA
Canadian Wood Frame Construction, Central Mortgage and Housing Corporation, Ontario, Canada
Home Renovations, Francis D.K. Ching & Dale E. Miller, Van Nostrand Reinhold, New Jersey, USA

Disclaimer

The author has made every effort to insure that all instructions given in this book are accurate and safe, but cannot accept liability for any injury, damage, or loss to either person or property—whether direct or consequential—resulting from the use of these plans and instructions. The author, however, will be grateful for any information that will assist him in improving the clarity and use of these instructions.

Jobsite Safety Provision

The author's instructions as written in the book, carried out on the construction site, shall not relieve the owner, or General Contractor of its obligations, duties, and responsibilities, including, but not limited to construction means, methods, sequence, techniques, or procedures, necessary for performing, superintending, and coordinating the Work in accordance with the plans, diagrams, and text in the book, and any health or safety procedures required by any regulatory agency. The author has no authority to exercise any control over any construction contractor or its employees in connection with their work or any health or safety programs or procedures. The owner agrees that he or his General Contractor shall be solely responsible for jobsite safety, and warrants that this intent shall be carried out in any contract he has with the General Contractor. The owner also agrees that the Author and the Publisher shall be indemnified by the General Contractor, if any, and shall be made additional insureds under the General Contractor's Policies of General Liability Insurance.

Full-Size Construction Plans are available from the author – please send e-mail or request To: John Hicks, P.O. Box 75, Keswick, ONT L4P 3E1, Canada jsh@interhop.net

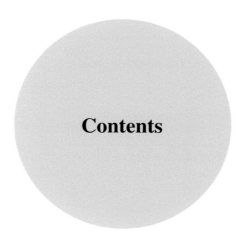

Contents

About the Author

John Stephen Hicksis a senior Landscape Architect specializing in Provincial Park Site Planning and environmental assessment. His spare time is spent managing his own forest and wildlife area, along with studying the surface of the Sun in his private observatory on his home property. Through the design and construction of his own observatory, he developed a complete package of construction plans which he has marketed worldwide.

Guest speaker at astronomy conventions in Ontario, Canada, John has for many years promoted the idea of building your own observatory, refining the process to the steps outlined in this book. He also designs and builds his own refracting and reflecting telescopes, again with plans to follow.

His special interest in astronomy is in the realm of solar observing and photography. For 25 years he has employed various Hydrogen alpha filters to examine the solar chromosphere, all the while improving photographic techniques toward capturing what the eye actually sees. He photographs flares, activity areas, and prominence dancing on the edge of the Sun. His astronomical presentations to astronomers are split between solar H-alpha techniques and observatory design and construction. He welcomes any inquiry on either subject, always interested in assisting people involved in astronomy.

Part I
The Roll-Off Roof Observatory

Chapter 1

The Benefits of a Permanent Observatory

Any astronomer familiar with setting up an equatorial telescope will realize the time required to level, polar align, and prepare an instrument for an observing session. In the case of a non-computerized system, the tasks involved with centering Polaris with its required off-set for the North Celestial Pole is daunting enough night after night. Even with a computer-assisted "scope," set-up time still involves the usual lugging of equipment out-of-doors from either residence or vehicle (although polar alignment is greatly reduced with computer alignment hardware). Final assembly can still stretch patience with the attachment of battery, dew heaters, and a myriad of wires connecting all the apparatus. Additional to all of this, many observers still have to carry out and assemble an observing table complete with sky charts, red light, lens case, camera and film. After completing this Herculean effort, particularly in northern latitudes, an astronomer usually begins to feel cold and exhausted while a degree of anxiety increases to finally use the instrument. This is often the prelude to damaging equipment or injury through acting too hastily with impatience. Repetition of such set-up experiences eventually discourages most observers who eventually reduce the frequency of their observing sessions, or trade the heavy equipment for lighter instruments with less aperture. The "lightening up" process works opposite to the usual "aperture-fever" affliction that burdens most astronomers with greater diameter lenses and mirrors and their subsequent weight increase. Under normal circumstances, amateur astronomers also find themselves observing out in the wind, in the cold, and eventually using an instrument that is covered with dew. In order to eliminate the majority of these unwanted effects, one really needs a permanent observatory. Simple forms of observatories are available, but almost of them also require a set-up time, offer little weather protection, and are not quite as "portable" as advertised. The primary decision involved with observatory design rests between choosing either a domed-type or a roll-off roof type structure. Both have distinct advantages however, depending on your personal observing needs—including the requirement for an all-sky view, protection from the environment, and degree of privacy. There are other design options for simple observatories such as the "clam shell roof," the new cylindrical domes, and various types of shelters or housings that roll away to expose the telescope. Although these may be simpler to construct, they most often expose the observer to the elements, and are more difficult to weather-seal when not in use.

© Springer-Verlag New York 2016

J.S. Hicks, *Building a Roll-Off Roof or Dome Observatory*, The Patrick Moore Practical Astronomy Series, DOI 10.1007/978-1-4939-3011-1_1

Pros and Cons: The Dome Versus the Roll-Off Roof

Many owners of the roll-off roof type prefer an all-sky view, and are willing to tolerate the residual effects of wind, cold, and less control over light pollution (without the benefit of being able to select specific sky segments as with the dome slot). They also may be interested in hosting large groups, which of course demands the more spacious accommodation offered by the roll-off model. The person demanding a high degree of privacy in his viewing may appreciate the canopy provided by the dome, although the side walls of a roll-off type observatory still afford a reasonable degree of privacy. It is hard to concentrate on solar viewing for example when a host of neighbors watch you set up and enjoy your fumbles. Solitude is important for concentration and speculation. Admittedly, the roll-off roof offers a substantial improvement over just an open observing site, while the domed observatory may further reduce most annoyances, achieving a completely sheltered structure. The crucial decision to make in selecting either is one of sky view. If you want to see the whole sky dome at once, the roll-off is the better choice (Fig. 1.1).

In addition, the roll-off roof type quickly cools down to ambient temperature with the entire roof rolled off and the instrument(s) entirely exposed to the open sky. Fast cool down is not as easily attained with a dome-type structure.

However, on the other hand one need only step outside to see the heavens, and concentrate on a portion of it inside. Cost factors and degree of skill enter into the equation also. The roll-off will be simpler to build (wood construction and no curved sections) and less expensive in parts and labor. But in terms of durability, the all-metal dome will outlast it.

Fig. 1.1 The advantage of a completely open observatory—the glorious night sky dome (Courtesy of Jeff Pettitt)

Fig. 1.2 The author's domed observatory—while the sky is largely obscured except through the observing slot, its advantages are largely shelter from the wind and to some degree, the sun (Photo by John Hicks)

Identification of the dome as the "symbolic" structure used by astronomers may also be an asset to someone who wants to advertise to the community that their hobby is astronomy (Fig. 1.2).

On the other hand, one may want to maintain a lower profile in high crime areas, preferring to "hide" the facility as a garden shed. In essence, the choice is dependent on many factors, including site constraints and budget, along with the particular objectives and skills of the observer.

Roll-Off Roof Variations: The Sky Is the Limit

The observatory I designed for Don Trombino in Florida, fulfills both astronomical and landscape functions. With its exquisitely finished interior, and practical outdoor patio under the gantry, this observatory stands out prominently.

The owner, the late Don Trombino, was so proud of his achievement, that he spent almost all his waking hours either inside it or under the patio. He further extended the observatory feature out into the garden with a stone paver walkway leading across the yard terminating with a sundial monument. The floor under the roof gantry was also set in stone pavers and the underside of the gantry "ceiling" covered in a prefab wooden lattice. When not solar observing, Don spent many hours on the patio, examining the results of his photography, or writing. He dedicated the structure "The Davis Memorial Observatory" and symbolized the dedication with various artifacts and historical items placed in the garden and on walls of the structure (Figs. 1.3 and 1.4).

Once and a while certain observatories stand out as truly professional structures, finished to the point excellence. Such a model is Mike Hood's observatory, complete with outside porch under the gantry, featuring a door on the gable end. Mike has put extra effort into tapering the hip roof back from the gable ends, adding a small "cottage look" to his observatory. Very tastefully finished, it has an interior just as spectacular. His structure is long enough to hold a complete control room with desks, cupboards, an air-conditioning unit, and a window. Overall the control room has the appearance of a high-tech whiteroom, temperature-controlled and very well designed. Apparently the observatory was from an original model by "Backyard Observatories" (Figs. 1.5 and 1.6).

Fig. 1.3 Close-up "Davis Memorial Observatory" with its patio garden under the gantry (Photo from the collection of John Hicks)

Fig. 1.4 "Davis Memorial Observatory" with walkways, sundial, and landscaping (Photo from the collection of John Hicks)

Gerald Dyck's roll-off roof observatory in Massachusetts presents a compact, attractive addition to his yard. The roof line is particularly well-designed with a skirt that extends down over the walls to keep out insects, and the elements. Note the use of an exhaust fan on the gable (Fig. 1.7).

Fig. 1.5 "Mike Hood Observatory"—a truly well-finished roll-off observatory (Courtesy of Mike Hood)

Fig. 1.6 Mike Hood's Control Room complete with desks, cabinets and air conditioning (Courtesy of Mike Hood)

Fig. 1.7 Gerald Dyck's Compact Observatory, well-protected from the elements (Courtesy of Gerald Dyck)

The wall height is also kept lower, presumably due to the roof skirt which replaces a portion of it below the normal soffit level. This allows for more accessible horizon-level viewing as the photo below illustrates. The telescope shown can reach lower elevations than most, swinging even further down than the position shown. The Dyck's prefer to utilize telescopes on tripods rather than on a fixed pier. Although quite suitable for alt-azimuth mounts such as a Dobsonian cradle (shown), their future plans will most certainly involve a fixed pier with an equatorial mount (Fig. 1.8).

Although the entry door is lower than a full height door, it is made more accessible by the fact that the observatory is raised off the ground considerably. Such an arrangement allows the operator to step up into the structure rather than stoop to get into it at more normal foundation levels. I used this technique myself on my first observatory which had only 4 ft high walls. The increased height of the floor off the ground also prevents skunks, squirrels, possums and groundhogs from seeking refuge permanently under it. There is little protection from wind or the elements with so high a crawl space underneath. It also allows alterations in wiring underneath or the addition of insulation under the floor. In crawl spaces like this, it is wise to line the ground surface with landscape fabric (two layers minimum), covering the entire area underneath with 4 in. of 3/4″ crushed gravel. This treatment prevents weeds, and discourages animals with its sharp edges of gravel. It also has an attractive, clean look underneath which prevents excess moisture, moss etc., from accumulating in the shaded environment (Fig. 1.9).

Dave Petherick of Ontario, Canada has built a well-landscaped observatory on a typical sub-urban lot. He has incorporated a beautiful deck complete with trellis as an integral part of his observatory design. In fact, the deck is an extension of the observatory which allows for a large area for entertaining, barbequing etc. The trellis appears to be an extension of the gantry which transforms it into a "landscape feature" thereby creating a dual function for the observatory—both astronomy and gardening. It also serves to "hide" the true function of the gantry which would lessen the footprint area of the structure, the gantry portion appearing more like a garden trellis. Complete with shutters on

Fig. 1.8 Gerald Dyck's low observatory wall design with Dobsonian telescope inside (Courtesy of Gerald Dyck)

Fig. 1.9 Gerald Dyck's Roll-Off Observatory showing open roof with low door (Courtesy of Gerald Dyck)

"mock windows," the observatory looks just like any other well-designed garden shed. This is the look I'm recommending builders aim for. Again, notice the gable fan which can be used prior to an observing session to evacuate hot air from the interior, bringing the observatory and contents closer to ambient temperature. Dave constructed a corner dark room for his computer, the monitor shown

Fig. 1.10 Petherick Roll-Off Observatory showing gantry transformed into trellis (Courtesy of Dave Petherick)

Fig. 1.11 Petherick Roll-Off Observatory showing deck and surrounding landscaping (Courtesy of Dave Petherick)

temporarily outside on a table. Overall, a very handsome design, well thought out, and an attractive addition to any rear yard (Figs. 1.10, 1.11, and 1.12).

Of the various types of Roll-Off Roof observatories I have seen, Guy Boily's log frame observatory wins first place for unique structural components. Only the roof appears to be constructed with traditional framing materials, finished in an metal or vinyl ribbed siding. The logs forming the four

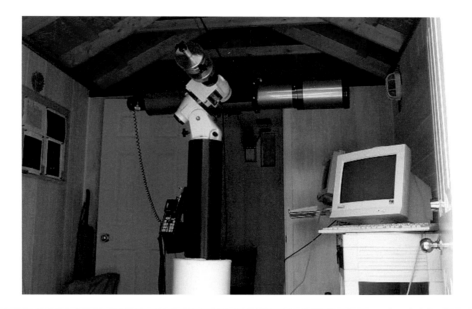

Fig. 1.12 Inside Petherick Roll-Off Observatory (Courtesy of Dave Petherick)

walls had to be carefully chain-sawed to lap one over the other at the corners, and further grooved along their lengths to fit comfortably together, avoiding the normal "chinking" process to seal off draughts. Looking at the walls from the inside there appears to be no air gaps whatsoever, requiring some precision in fabrication. Bracing the gantry appears to be accomplished by burying poles on an angle resisting the force of the roof being pushed out over rails toward them. The top log, which would be the caster rail nailed to the top plate in our model, is a continuous log pole in Guy Boily's model which simply extends out over the entire run of the gantry (something we cannot do with traditional lumber with its 16 ft maximum length). The castor track appears to be a steel angle fixed to the top log throughout, and the casters running on it fixed to the roof framing. Its refreshing to see the use of unique materials and a departure from the usual "balloon frame" construction (Figs. 1.13, 1.14, 1.15, 1.16, and 1.17).

Observatory Kits and Complete Observatory Fabricators

For those who do not wish to construct an observatory themselves, or who need it in a kit form, there are suppliers who will manufacture and install complete observatories. Often included in the their list of services are options for rolling left or rolling right variations, material lists, step-by-step construction advice, steel pier plans, and even maintenance hints.

Some of the sizes offered range as high as 10 ft × 14 ft—nearing the size of the model featured in this book. For the more senior astronomer, or one who is not familiar with using hand tools, or understanding diagrams, it is a definite option. These models possess light construction techniques throughout, and sturdy engineering where it counts. Overall the customers seem quite satisfied and proud of their observatories. Although this book is directed toward constructing an observatory from plans of my own design, I still feel it is fitting to offer options that might otherwise be suitable for

Fig. 1.13 Log frame observatory with roof in closed position (Courtesy of Guy Boily)

Fig. 1.14 Log frame observatory with roof in open position (Courtesy of Guy Boily)

those that do not have the resources to build it. And, by coincidence, a close associate of mine has such an observatory which I would like to illustrate. The main attraction with this particular model is the use of garage door track and rollers, which for a small roof load seem to operate satisfactorily. This arrangement omits the requirement to make a track and purchase the V-groove casters outlined

Fig. 1.15 Log frame observatory showing log walls with corner overlap (Courtesy of Guy Boily)

Fig. 1.16 Log frame observatory looking inside (Courtesy of Guy Boily)

Fig. 1.17 Log frame observatory gantry (Courtesy of Guy Boily)

Fig. 1.18 "The Ussher Observatory"—a prefabricated model by "Sky Shed." (Courtesy of Terry Ussher)

in this book. However, for a larger roof section, such as the size that our model requires, I would highly recommend the more sturdy assembly that I have detailed. Overall, the prefabricated structures look like a normal garden shed, and are complete with planked siding, steel roof, door and optional window (Figs. 1.18, 1.19, 1.20, and 1.21).

Fig. 1.19 "The Ussher Observatory"—close-up of structure with roof closed (Courtesy of Terry Ussher)

Fig. 1.20 "The Ussher Observatory"—roof interior showing rafters and no collar ties (Courtesy of Terry Ussher)

Fig. 1.21 "The Ussher Observatory"—close-up of garage rollers and track (Courtesy of Terry Ussher)

Other Types of Roll-Off Roof Observatories

Some roll-off roof systems are designed to roll off in two halves, in opposing directions which requires two separate gantries and two sets of tracks. This arrangement lessens the roof load on each separate gantry, but requires more workmanship in building and aligning the two gantries. It also introduces the prospect of misalignment of the extra gantry due to settling, wind storms, etc. Several techniques have been used to solve the problems of roof flashing where the two halves come together upon closing. A simple solution to securing the roofs together once shut, is the use of large wood clamps where they meet. An additional clamp would have to be placed at either end of the observatory (Figs. 1.22, 1.23, and 1.24).

The heart of this ingenious roof system is the four wheel beam assemblies which are made of 3″ × 3″ steel box tubing. The box tubing was cut open to house two wheels on each length. Each side of the 10 ft roofs held one box tube complete with its two rollers. The wheels run in steel angle just as we are proposing in our design (Fig. 1.25).

Tabs welded onto the top of the box tubing allowed attachment to the roof trusses. This design allowed the roof to "hug" the top plate very closely, almost eliminating the usual air gap between roof and walls. An extra bonus of such an arrangement is the ease of keeping the track and rollers clean. Overall, a very efficient design and worthy of copying by those with milling and welding skills (Fig. 1.26).

Weather-proofing the rolling roofs usually requires a "Z" type flashing on one half with an angle strip on the other half which the top flange of the "Z" flashing slides over. The design and braking of these flashings are best left to a roofing or heating contractor, who should be consulted if you wish to go this route (Figs. 1.27 and 1.28).

With the increasing use of computers and CCD cameras, it is very advantageous to incorporate a "warm room" into the observatory structure to keep the computer warmer than ambient air in cool seasons. It also serves as a dark room to house the astronomer preferably at a small desk with a comfortable chair. A particularly unique variation in possible designs is placing the observing part of the structure lower than a "warm" room section. This arrangement allows for a smaller roll-off roof por-

Fig. 1.22 "Rocky Plains Observatory"—a twin roll-off roof observatory showing open roofs (Courtesy of Bob Luffel)

Fig. 1.23 "Rocky Plains Observatory"—showing roofs closed and the gantry braces (Courtesy of Bob Luffel)

Fig. 1.24 Roof Clamps hold the two halves together when the observatory is not in use (Courtesy of Bob Luffel)

Fig. 1.25 "Rocky Plains Observatory" roller system—the wheels are V-groove, same as suggested in our plan but Bob has ingeniously enclosed them in a steel box beam—each roof section carries its own two sets of box beams enclosing the wheels (Courtesy of Bob Luffel)

Fig. 1.26 "Rocky Plains Observator" inside track—illustrating how snugly the roof joists and thus the roof hugs the top of the observatory wall (Courtesy of Bob Luffel)

Fig. 1.27 Z-flashing on open roofs—"Rocky Plains Observatory." (Courtesy of Bob Luffel)

Fig. 1.28   Z-flashing on closed roofs—"Rocky Plains Observatory." (Courtesy of Bob Luffel)

Fig. 1.29   "Ballauer Observatory"—illustrating higher warm room section, lower roll-off roof, and attached open storage area (Courtesy of Jay Ballauer)

tion, which is lighter and shorter than a design which tries to put both uses under one long rolling roof. This is a very effective solution accommodating both uses. The rolling roof rolls off in the opposite direction to the <u>fixed</u> "warm room" roof, and simply "tucks" under the fixed warm room roof gable when closed shown in the Ballauer Observatory (Figs. 1.29 and 1.30).

Another innovative variation resembles an "A-frame." When rolled fully apart the two halves create an open-sky viewing area. The telescope rests on a pier in a somewhat "lofty" position well above

Fig. 1.30 "Ballauer Observatory"—illustrating how roll-off roof tucks under the higher warm room roof above it. Note casters and the large air gap between roof and walls (Courtesy of Jay Ballauer)

the level of the observer's head—a good position for a refracting telescope. Because of its height, the telescope can swing to most areas in the sky being high-up towards the "peak" of the two roof halves.

When finished for the night, the owner simply points the telescope east-west horizontally, pushing the two halves of the structure together, then exits via the undersized door. Notice how the overhang on the north end of the observatory will clear the horizontal tube when in a closed position. The taller (north) half blocks an area of the circumpolar region under Polaris where little observing is done. Each half moves on six casters running on 20 ft wooden rails set 10 ft apart. The owner finds that sometimes it is not necessary to roll both halves apart: one side can remain stationary to block the wind while the other moves to an advantageous position for viewing.

When closed, this observatory measures only 10 ft × 10 ft, just fulfilling the maximum area size under building code exemption. The observatory eliminates the effort of constructing a roll-off roof track and gantry, although I suspect in a harsh climatic zone, a foot of snow could provide a real nuisance, requiring shoveling the rails and platform clean. Once left for any period of time encrusted snow would create a larger problem (Figs. 1.31, 1.32, 1.33, and 1.34).

Observing Sheds

One solution to eliminating the roll-off roof and gantry entirely is to construct a normal garden shed in either hip roof or barn roof format, adding a small enclosure to it at the south end. This is constructed large enough to allow full instrument swing and room for the observer. The roof on this addition is simply a normal shed-style roof, hinged to swing upward with appropriate braces to hold it in position vertically. It could also be designed to slide off or lift off much like a cabin hatch on a sailboat. The roof will have to be constructed of light-weight materials if it is a lift-off arrangement. Ideally, the telescope, mount and pier are fixed in place under the "hatch" and polar aligned. This can

Closed Roll-Apart Observatory- (During Construction)

Fig. 1.31 "A-frame" style observatory built in two rolling halves showing observatory closed (Courtesy of Greg Mort)

Fig. 1.32 The owner seated inside the "A"-frame observatory with the two halves open and a cloth screen for wind protection (Courtesy of Greg Mort)

Observatory floor & pier construction

Fig. 1.33 Tall brick pier designed to hold a large refractor just below the roof peak such that it will clear the lower roof half. Note the size of the cement footing just visible under the pier and the stepped nature of the brick column (Courtesy of Greg Mort)

partially framed roll-apart observatory-showing casters

Fig. 1.34 Sturdy observatory foundation showing the six castors on each side. Also note the simple construction which eliminates fabricating roof trusses. The walls are the trusses (Courtesy of Greg Mort)

Fig. 1.35 Barn-Shed style Observatory "Big Woodchuck Solar Observatory" showing the roof hatch open ready for observing (Courtesy of Larry McHenry)

be accomplished with a suitable pier if there is room for it, or a tripod that rests upon small pedestals under each tripod leg. To assure that polar-alignment is maintained, the small pedestals must be impressed in some way to accept the legs in the same orientation each viewing session. Conversely, a telescope on a tripod could be hefted from the shed portion into the addition and polar-aligned each session, but this defeats the purpose of an "instant" viewing session afforded by a permanent, fixed situation. Its drawbacks include little or no view of Polaris and the northern celestial sphere, plus the difficulty to keep a tripod polar-aligned unless a permanent pier is poured in position. A permanent, fixed pier is the best solution which takes us back to the roll-off roof design (Fig. 1.35).

Clam-shell Roof observatories are worth investigating, even though their roof(s) do not roll-off. Built on a low box-type wall, they offer the maximum sky view, almost completely exposing the astronomer to the elements. Most are designed such that the operator must step over the low wall into the structure, which in the dark becomes a hazard for visitors who have to be guided over it. Occasionally, a builder includes a low hinged entry door as in the model shown below. The roof halves are each counterweighted with heavy cement blocks or thick steel plates on arms which extend outward from the roof "gables." These "arms" must be constructed such that they will not hit the ground surface as they rotate when the roof is swung open. The "arms" must also be of sturdy construction to carry the heavy counterweights. Often it will be necessary to dig a channel alongside the observatory walls to allow for the swing of the counterweights, which appears in Fig. 1.36 (North York Astronomical Association observatory at the "Bog," designed by Andreas Gada, north-west of Toronto). Notice the supports on this model holding the heavy roof sections once opened, to ease the stress placed upon the hinges. The other model shown with particle board exterior has no such support but is of lighter construction (Fig. 1.37)

The main difficulty in building this type of structure is balancing the roof sections with the counterweights (which can only be done with trial and error procedures), and the construction of the triangular roof sections. For all the trouble in constructing this type of observatory, it does not offer the protection nor convenience of the roll-off roof model. A major complaint with the structure is leakage through the roofs at the seam, or at the junction of roofs with the low walls (Figs. 1.36, 1.37, and 1.38).

Fig. 1.36 North York Astronomical Association Observatory at the "Bog" with roofs open showing the small entry door and excavated channel alongside wall for the swing of the counter-weights (Courtesy of Danny Driscoll and NYAA)

Fig. 1.37 Paul Smith's Clam shell Observatory closed showing slope on split roofs (From the collection of John Hicks)

Fig. 1.38 Paul Smith's (1995) Clam Shell Observatory open showing 10 in. reflector. Note that the north roof half swings only vertical affording an excellent shelf for equipment cases, lenses etc (From the collection of John Hicks)

Art Whipple's Roll-Off-Observatory

In the realm of observatory design, someone eventually comes up with a structure that is so unique it represents a break-through in operation and begs the question "why didn't I think of that"? Art Whipple of Maryland USA has designed an entire observatory that rolls away from its telescope. An experienced solar observer, Art needed an open-air station to achieve high-resolution solar imaging that was not influenced by the heating of an observatory building. The solution to improving seeing was to roll the observatory housing the telescope away from it, to the north, where convection currents from the sun-heated roof rose away from the path of observation. Art's solar telescope of his own design, was too heavy (45 kg) to remove from the structure, and carrying it to a distant pier was not possible. A movable observatory exposing the telescope was the solution (Fig. 1.39, 1.40, 1.41, and 1.42).

The cypress hedge surrounding both telescope and observatory offers wind protection, preventing any buffeting of the solar telescope. To ensure that no heat is trapped within the confines of the encircling cypress hedge, the lower stem areas are kept pruned and open, allowing air to circulate underneath. The observatory rolls on four grooved steel wheels that run on a pipe track, much like a railway car runs on its track, complete with small ties imbedded in the lawn. Weighing 180 kg, the entire observatory rolls forward and backward with ease. The base of the observatory is a heavy platform structure containing a slot to accommodate the pier and telescope in its stow position to the north. A removable section of platform covers the slot when the scope is stowed to keep out insects and provide weather-proofing (Fig. 1.40).

A pair of gas springs counter-balance the weight of the observatory as it is opened (much like a hatchback on a vehicle). Hinged at its north end, the 90 kg observatory cover can easily be lifted by one hand (Figs. 1.41, 1.42, and 1.43).

The scope contains an un-aluminized 35 cm paraboidal primary mirror at a focal ration of f/4.6. Supported by an 18 point flotation cell, the mirror suffers little or no degradation of its surface figure.

Fig. 1.39 Art Whipple's roll-off observatory in its open position, tucked north of the telescope within an enclosure of Leyland cypress (Courtesy of Art Whipple)

Fig. 1.40 The telescope under the roof of the observatory in its stow position. The removable section of platform has been inserted (Courtesy of Art Whipple)

Fig. 1.41 View of the open observatory in the distance, moved away from the telescope. Once the observatory is rolled fully back over the scope, and the removable section fitted, the upper portion is lowered offering a dark room for the computer and video monitor. A small door on the north end provides access inside to the electronics (Courtesy of Art Whipple)

Fig. 1.42 Once fully rolled back away from the 'scope, the interior of the observatory is evident. Art provided a roof overhang on the north end visible in Fig. 1.39 that provides welcome shade while imaging (Courtesy of Art Whipple)

Fig. 1.43 The solar telescope construction—an open Newtonian design with the camera position at the prime focus position (Courtesy of Art Whipple)

In addition to this the mirror is ventilated by three fans running at low speed of 600 rpm. The benefit of this design is that the camera (a DMK 31 AU03) is placed at prime focus not at secondary focus as in a Newtonian, in this case yielding the full f/4.6 primary beam. To minimize obstructions, the camera is also supported by a spider made of 0.5 mm stainless steel that offers a stiff, easily adjustable, support.

Art's solar white light photography is outstanding, resolving solar granulation completely, and delving right down into the fine filigree of sunspots. The design of his structure can be considered as much a part of his imaging skills as the design of the telescope itself—truly a magnificent idea.

The "Clapp Trapp"

The simplest of designs involves a lightweight, transportable "hut" which assembles in panels with a roof "lid." The panels are hinged together with upper portions that can be hinged down individually to allow viewing near the horizon. The roof "lid," sits on top of the vertical side-panels with its own small hinged panel which can be swung open and back to create a "slot" for viewing. Overall the design was very simple yet effective, and easy to put up in 10 min. It was affectionately named the "Clapp-Trapp" after its creator, Douglas Clapp, and found use throughout Ontario with stargazers who wanted a quick, temporary shelter for observing. Large enough to house a single astronomer comfortably, the walls held pouches for accessories which could be reached with a turn of the body inside. Fully erected, each section stood 6 ft high, and 4 ft wide (composed of 2 ft×2 ft sections

Fig. 1.44 The first model of the "Clapp Trapp" showing the numbered wall panels, with some wall sections swung down, and the roof still in place. The purpose of the wall panels that swing down was to observe objects nearer to the horizon. Not all wall sections have panels that swing out to maintain stability. Too many panels swung down would leave the structure unsupported

hinged together). A tarpaulin placed on the ground before assembly kept out moisture, snow, and insects. The entire structure assembled weighed in at 125 lb. Requiring only about three evenings to construct, it could be built from 4-1/2 sheets of 1/4 in. plywood for less than $200.00. Plans for the model are still available from the author, now known as "The Super Portable Observatory" (Figs. 1.44, 1.45, 1.46, and 1.47).

Fig. 1.45 The "Clapp Trapp" was easily folded up into a 14 in. high stack of folded panels measuring 2 ft × 4 ft and weighing only 125 lb. It fit into most larger vehicle trunks (Courtesy of Doug Clapp)

Fig. 1.46 The "Clapp Trapp" walls were hinged in pairs allowing them to be self-supporting during assembly. They hooked together to form an octagonal structure that supported the "floating" roof. Note the smart use of "pouches" to hold accessories on the inside walls (Courtesy of Doug Clapp)

Fig. 1.47 The "Clapp Trapp" roof assembly showing the folding observing hatch and its "prop-up stick". To follow the motion of the stars, the roof was simply lifted up slightly from inside and rotated into proper position (it weighed only 25 lb) (Courtesy of Doug Clapp)

Chapter 2

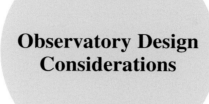

Observatory Design Considerations

Size and Internal Space Required

Several criteria combine to create the minimum size requirements for your observatory. These include the height of the astronomer first and foremost. The focal length of the telescope combined with its type of mounting are primary factors to consider next. The optical configurations of existing and future instruments planned for are equally important whether refractor, Newtonian, or Schmidt Cassegrain models. The walls surrounding the observatory are usually designed to allow the optical axis of the telescope to clear them when it is in the horizontal viewing position. Since this is also the usual storage position, the walls will have to be higher to allow the rolling roof to completely clear it when rolled shut. This is a necessity since the gable ends or roof rafters would strike the telescope on closing. At any rate, seldom does one observe on the horizon, or even a degree or two above it due to light pollution or landscape obstructions, so the need for horizontal viewing is largely unnecessary (Figs. 2.1 and 2.2).

The German Equatorial Mount, possibly the most ordinary configuration, is the most demanding to accommodate since its swing and roll-over positions require an extra space tolerance beyond most other mountings. For refractors, a typical allowance between eyepiece and walls might amount to about a meter—still wide enough to allow for some crouching on the part of the observer. Newtonian reflectors, which place the eyepiece off the optical axis on the telescope tube side need less allowance for the observer viewing objects at the Zenith, since he/she is likely standing for such an observation. *Clearances* as low as 1/2 m are suitable in this position, provided the observer is standing and not bent over. In the case of Newtonian-type telescopes, the mount is usually designed to place the eyepiece at eye level, and is dependent upon the height of the observer (Fig. 2.3).

Lower-angled Newtonians often need a bent or seated position on the part of the observer, and possibly the accommodation of an observing chair, which will require more clearance.

Refractors on German Equatorials present the observer with the same awkward positions.

The difference is largely due to the fact that the observer looks down the optical axis of the refractor which places him/her at the very end of the telescope, really squeezing the observer against the walls or even the floor in certain viewing orientations.

© Springer-Verlag New York 2016
J.S. Hicks, *Building a Roll-Off Roof or Dome Observatory*, The Patrick Moore
Practical Astronomy Series, DOI 10.1007/978-1-4939-3011-1_2

Fig. 2.1 Mid-construction test: Refractor fits under closed roof when rotated completely horizontal. If calculations had been incorrect, the pier or cement base pedestal would have to be shortened (From the collection of John Hicks)

Schmidt-Cassegrain type reflectors, although compact, are often mounted on a fork and wedge assembly which is adjusted to the observer's latitude. This arrangement offsets the instrument from the pier and is usually best arranged so that the offset is to the south (mainly because the majority of deep sky objects locate in the southern sky). This allows more space for the observer to sit or bend down on the north side of the pier, but constricts space on the south side. The zenith position of the telescope usually places the eyepiece within the confines of the forks, (held in a diagonal), and the telescope pointing vertically. Often this position is more easily accessed on the other side of the mount, with the diagonal rotated toward the south wall where space is more limited. Respecting all these criteria when planning will produce a comfortable volume for you to observe within, but neglects the requirement for visitor space.

If your interests include accommodating groups, the above measurements will have to be increased to provide ample space.

Fig. 2.2 Completed "Davis Memorial Observatory" with refractor at rest position ready for roof closure (From the collection of John Hicks)

Fig. 2.3 Don Trombino imaging the Sun with ample wall clearance behind him. With this much Instrumentation, it is essential to have a good working distance (Courtesy of Jeff Pettitt)

Tables have been created listing instrument type, focal ratio, telescope tube length, mount height, and floor area required. Actual measurements of your instrument on your pier of choice in all its various swing positions is actually required. Nothing substitutes for a well-planned analysis of your present or future instrument. It is always best to err on the design toward oversize because your instrument may change with time. The expenses involved with a slight increase in size are not excessive if you are self-constructing it. Build your observatory with the prospect of expanding it someday. The requirement for a computer room/warm room is better accommodated now than having to add onto the structure later. The design which involves a higher roofed, permanent warm room with a lower roll-off roof section is the way to go. I would not advise the lengthening of the roll-off roof section as that will increase its weight and load on the track, requiring extra casters and more force to move it on and off.

An alternative arrangement toward establishing a warm room using the model exhibited in this book involves the addition of a lower-roofed extension under the gantry in the north end. With its own permanent roof, insulated wall between, door, and a thermal window between the rooms, it will function as an ideal warm room without disturbing the gantry or compromising the roll-off feature. Although the roof will be low, working on a computer is normally more comfortable in the seated position, so complete standing room clearance is not necessary.

Advantages of a Roll-Off Roof Observatory

The decision to build a roll-off roof type observatory affords the following benefits:

(a) the telescope, mount, and drive components are protected from the elements when not in use.
(b) instrument cool-down time is fast with the entire roof rolled away, and ambient air temperature is reached quickly.
(c) some equipment in use can be protected from dew if the roof is partially rolled off.
(d) wind and cold protection is partially afforded from ground level currents with adequate side walls.
(e) offers a permanent storage area for charts, books, and even equipment.
(f) offers a permanent electrical connection, covered and water-tight.
(g) offers a safe lock-up area if wired to a house alarm system.
(h) increased security through its shed-like appearance, masquerading as a normal out-building (unlike the domed type which portrays a scientific use with its associated expensive instrumentation contained within).
(i) the option to use it for general storage purposes if the hobby loses its appeal.
(j) opportunity for work station, greater visitor accommodation and group sessions (Fig. 2.4).
(k) generally easier to accept on a re-sale of the property because it masks as a shed.
(l) easier to construct than a domed observatory.
(m) usually less expensive to build than a domed observatory (aluminum domes are higher in price).
(n) no special tools required (except for the welding of track and dome base ring).

Questions to Ask Before You Begin

1. Do you have sufficient space to build an observatory? (the roll-off roof type requires more space than a domed observatory requiring a roll-off gantry frame).
2. Are you planning an outdoor patio? The roll-off roof observatory can be utilized as an outdoor sitting area if the gantry portion is improved with a sun trellis just under the gantry frame.

Fig. 2.4 Interior of a domed observatory leaves little space for a work station or room for more than three visitors due to curved walls and the central pier (Photo by John Hicks)

A stone paver patio can be added underneath the gantry thereby disguising the whole observatory as a shed with patio attached.

3. Are you willing to exert the effort to build an observatory?

4. Does your observing program demand an all-sky view?

5. Does your local building code exempt structures less than 100 ft^2 in area from building permit requirements? (if so, it might be to your advantage to limit the observatory dimensions to 10 ft × 10 ft, although the building code may not exempt the extra square footage contributed by the gantry—a dome observatory requires less space).

6. Do you plan to host large groups of observers or star parties?

7. Will the roll-off roof model house your present or planned instrument? Beware that longer refractors require a large volume to swing through all viewing positions.

8. Can you orient the observatory to fulfil all the necessary sideyard requirements of your local Zoning By-Law? This could also include rear yard set-back, side yard set-back, and lot coverage (the % of the lot covered by buildings). The By-law might also forbid the construction and use of an "accessory building" without a "main use," meaning that you cannot construct an observatory without a residence being there first, or at least simultaneously).

9. Assuming you have a partially treed lot, can you site the observatory with enough open sky opportunity? Will it require some tree removal? Note that some municipalities have stringent tree by-laws which require an exception to the by-law or a minor variance and a permit to remove. Careless removal could result in a fine.

10. Is the star Polaris accessible in your chosen location? (a necessity for polar-alignment with non-computerized telescopes).

11. Some counties/municipalities define the "permanence" of a structure as the limiting factor in exceptions to the local building code. Placing the structure on footings might qualify as a temporary-type building, whereas pouring a huge concrete (permanent) slab, may not. In all cases, describe your project as a garden shed observatory. Larger observatory structures will require

engineering approval and a building permit, whereas a smaller garden shed-style observatory will generally fall within the exception limits of the building permit requirements. Research the application of your town's building code, find out how your structure can be permitted, and then build it to suit the requirements of the code.

Overall Site Requirements

Polaris and the Southern Sky

A prime concern in siting the observatory will be the availability of Polaris and the southern sky. The star Polaris is essential for polar alignment of your instrument on the pier during observatory construction, and in years following for accurate tracking. In addition, since the southern sky has so many interesting objects to explore, its access is almost a prerequisite for siting an observatory. If foliage does block the view to Polaris, it can be selectively pruned for the few times that one does need access, although this may require permission from adjacent owners. It may also take some precision, requiring two people — one at the finder telescope eyepiece, and one distant with a pole saw.

Access

Considering the tools and materials that are required to build the observatory, access by vehicle is important, particularly when a cement truck is available to pour the slab and telescope pier. The use of a portable gas or tractor-driven cement mixer is quite acceptable but the provision of water, mix, and construction materials will require some slugging from a roadside vehicle to the observatory site. Also, vehicular access becomes a critical factor when planning future star parties, and along with a small parking area, makes the observatory all the more functional. A good road with a good base and sufficient gravel added will make the venture more feasible.

Electrical Service

The potential to hook up to electrical service is a factor to consider in the siting of an observatory, eliminating the need for an electric generator. In the future it will become increasingly obvious that electrical service is a real asset, providing reliable lighting, steady telescope drive control, heating for a warm-up hut, and possibly security alarm power. Poles carrying service are cheaper than burial of wire but obvious against the sky if observing low on the horizon. In addition, buried cable is very expensive. In some states and northern municipalities burial of wire is mandatory.

Elevation and Seeing

Second in importance is the elevation of the proposed site in relationship to the surrounding landscape. Obviously, the higher the chosen site is with respect to land around it, the more available is the southern horizon for viewing. Also, a higher site will tend to reduce air turbulence near the ground, particularly if the observatory is on a knoll or hill top. This is largely because rising, hot air from the surrounding landscape is confined somewhat to lower elevations. Every meter above the surrounding base plane will normally improve seeing.

Soils and Drainage Suitability for Footings

Several issues confront the observatory builder in addition to maximum sky access, and the technical constraints required for good astronomy—probably the most important is drainage. Make sure your chosen location isn't in a low area, or on a drainage course. Permanent wetness along with freeze-thaw cycles will destroy concrete footings and slabs faster than any other environmental effect. Footings may fracture and slabs will crack or flake with continued wetness and freezing. Even moderate dampness will flake the top surface of a cement floor once frigid weather settles in permanently. If your chosen site is wet and not on a drainage course, attempt to fill it in. Adding a good sandy soil mixture, tamp it firm or compact it by rolling it repeatedly with a heavy turf roller. You can alternatively let it sit over a year for the soil to compact under its own weight before any construction commences. On excavation, a wet sub-soil can be improved with the addition of plenty of 3/4″ crushed gravel around and underneath the sono-tube footings or under a cement slab if you choose to go that route. This allows air to circulate and dry out the interstitial spaces between gravel and soil, maintaining a dry column around and under the cement forms. This technique is usually mandatory in the construction trades whenever concrete is poured. If your problem of drainage is severe, you can also install sub-surface perforated tile drains around your proposed observatory site, ushering the water elsewhere. Just make sure it isn't onto your neighbor's property or you could face a violation under the Drainage Act. The tile drains should also be encased in a trench of 3/4″ crushed gravel on all sides so that the drain is surrounded, even on top. This will aid in percolation through the "pipe" and also in drying out when flows reduce. Over time this will seed in with a light covering of grass, which will hide its presence and not seriously affect its drainage characteristics (Fig. 2.5).

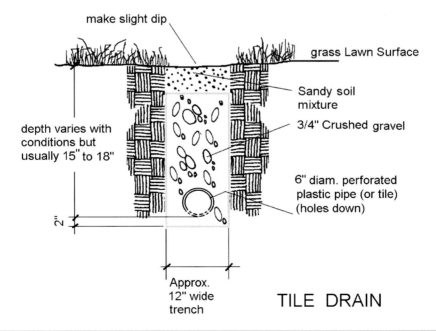

Fig. 2.5 Installation of tile drains will lower the water table around your site. Carefully directed away to a lower spot on your lot, they will transfer the moisture to a less critical area (Diagram by John Hicks)

Ground Surfaces, Turbulence and Light Pollution

Seeing also depends largely on the ground surface composition adjacent to and nearby the observatory. Bare soil, cultivated soil, hard surfaced areas (such as roads, parking lots, and house roofs) all have a deleterious effect on seeing due to heating during the day, and the retention of heat which radiates slowly off into the near-ground air column in the evening hours. Hence, the surrounding ground surfaces and land uses will affect the overall performance of your instrument and the practicality of your chosen observatory site. The asphalt and cement surfaces of roads and parking areas are the prime destroyers of good "seeing." Second to these are rooftop areas with their rising columns of hot turbulent air (refer to "emitted radiation" Fig. 2.6).

In the realm of solar observing a close water surface becomes an asset, and a coniferous forest type cover produces less transpiration than deciduous trees, creating less turbulence. These conditions are well met at the Big Bear Solar Observatory in Big Bear Lake, California. The observatory is actually built on a man-made peninsula out into the lake. The lake water stays cool so convection does not disturb seeing, and the smooth lake surface produces mainly a laminar airflow. The site is located on a mountain tarn surrounded by a heavy coniferous forest (Fig. 2.7).

Observatory sites to avoid include the following:

Airport runways (asphalt/cement surface, jet exhaust)
Large parking lots
Highway interchanges and major super highway easements
Heat-producing Industrial plants
Pits and Quarries (not yet rehabilitated)
Large roofs in your best line-of-sight

You should give priority to any "window of opportunity" for access to the southern skies. This may mean re-assessing your situation, and moving to another location within a site to utilize the most important segment of the skies.

At this point also, you must address the threat of distant light pollution: existing or potential. A line of trees, a forest edge, or a higher elevation between your site and a source of light pollution can be an asset. Although you may not be able to view right down to the horizon, the aesthetic advantage of

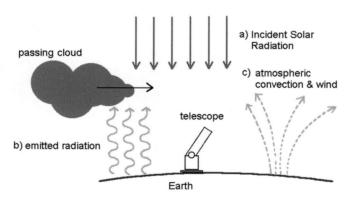

Surface heating and the normal balance of incoming and outgoing energy

Fig. 2.6 Diagram showing the effect of radiation, reflection, re-radiation and convection, on Earth's surface (Diagram modified from "Nearest Star" by Jay Pasachoff)[1]

Fig. 2.7 Big Bear Observatory on a man-made peninsula, Big Bear Lake, California, benefits from a laminar flow of air across a lake surface, high altitude, and a surrounding coniferous forest—all the necessary conditions to improve seeing (Photo by John Hicks)

a curtain of trees blocking a light source can be an improvement. The time and expense of building an observatory will become a major frustration if the viewing isn't near optimum upon completion. Consider the adjacent land zoning designation before you begin to predict the future lighting impact upon your site. Serious thought must be given toward all site requirements in order that your observatory project will be a successful venture (Fig. 2.8).

Zoning: The Surrounding Land Use

The zoning or land-use designation of properties in the immediate area should be of great concern to the observatory builder. Lands zoned for high density housing, particularly apartments and condos, often with flat expansive roofs and asphalt parking surfaces should influence your decision to build. Similarly, industrial or commercial zones on land which may now be vacant, constitute a real threat to your future use of the facility.

In this respect, you must consult the Official Plan (OP), and the Zoning By-law within your County, State, Regional Municipality or Town to ascertain what uses are allowed in each zone designation around your chosen site. Ask the Municipal or Town planner for zoning assistance, and explain the special requirements of your situation. Encourage the planner to predict trends in rezoning that are occurring, because these may have surfaced at public meetings held in the past for development proposals of the future. There may also be some restrictions that can be imposed upon a developer to reduce light pollution, parking area buffering (trees or berming) or no tree cutting etc., when proposals come to a public hearing and conditions are considered to reduce impact on the surrounding properties.

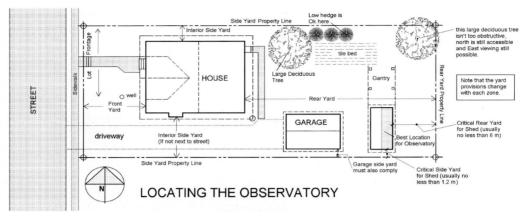

LOCATING THE OBSERVATORY

1. Determine North with a Compass
2. Determine the minimum sideyard and rear yard restrictions and measure in from the lot lines (string the bars together)
3. Position the observatory building to take advantage of south, south/west and west viewing opportunities if possible
4. Avoid water wells, water lines, septic beds and septic tanks
5. Have your local service provider from hydro, telephone and gas line companies locate all service lines for you.

Fig. 2.8 A typical suburban lot complete with garage showing the critical zoning provisions which change for each zone. Note the multitude of factors which direct the observatory location: side-yard minimums, rear yard minimum, proximity to tile field, well, trees, and the requirement for a northern alignment. Normally the rolled-off roof should go in a northerly direction (Diagram by John Hicks)

If a proposal is already launched, either before you begin, or while the observatory is under construction, consider appealing any Official Plan change or Zoning Amendment that might restrict your "continued use and enjoyment of the structure." Explain your requirement of reduced lighting to the council or committee during the public hearing.

It is better to attend this in person rather than sending a letter so that you may be able to speak about your project, the investment, and the scientific merit of your project. Your concern may constitute grounds for modification of an application for zone change, a consent, or a minor variance. It may even limit the developer's lighting of certain areas in order to reduce the impact upon you, an adjacent property owner. You have some bargaining power at this point because the developer will "bend" to conditions under the pressure of a refusal from council or the prospect of an appeal from a citizen holding his venture up.

Zoning: Limitations on Your Own Property

With regard to zoning, attention must be given at this point to the property upon which the observatory is to be built. Most zoning by-laws require a "main use" on the property (such as a dwelling) before an "accessory use" (such as an observatory) can be permitted. The implication of such zoning is that an accessory structure cannot precede a dwelling on the parcel, for reason that it could be used as an (unsatisfactory) residence.

An exemption from the by-law must be sought for the observatory if there is no residence on the parcel, through the mechanism of a minor variance. In this case, it will have to be launched by the owner of the parcel, and usually approved by a Committee of Adjustment.

A public hearing is required, and sufficient reason will have to be given to convince the committee that no improper use of the structure will take place. It is likely that a warm-up hut will be discouraged since it will constitute the very use the by-law is trying to prevent—that of a structure used as a residence.

I was myself trapped in a similar situation, my observatory constructed on a vacant parcel of land—contravening the "accessory use without a main use" statutory. I was forced to launch a rezoning application, relieving me from the by-law. The council questioned the possible use of the building as a domicile (home), and despite a heated argument over the fact that it was actually part of an instrument—and only its protective housing, produced no mercy on their part. The cost was excessive, and the hearing plus the 30 day appeal period put me behind schedule. There were no objections from surrounding neighbors, and although delayed, I eventually had my observatory. Oddly, years later, the rezoning protected me from light pollution from a cellular radio tower nearby which threatened my skies. I argued that I could enjoy "protection" under the by-law from light intrusion. I claimed by correspondence to the owners that they were "obstructing the application of the by-law." They didn't realize that "light intrusion" was not a condition of the by-law, and a tower was built further away. I felt pretty well satisfied over my achievement, being encumbered unfairly myself years earlier by an unnecessary restriction.

Ownership Versus Leasing

Ownership of the land is probably the most controversial item to be considered in the siting of an observatory. The merits of private ownership are obvious, but often the magnitude and expense of the project requires leasing. If leasing a portion of a parcel is possible it should be undertaken only with appreciation of future problems. The maximum lease duration in most areas for a parcel of land is 21 years less a day, without application for consent (formal land severance and deeding). In the case of an observatory, the lease should not be for any less time considering the effort and cost that is required to complete the observatory. In the case of a club leasing a parcel, it should be done after incorporation of the club, since this will save the directors from direct legal suit and allow the club to handle its assets in a legally accepted manner. If leasing is possible, the observatory functions must regard the sanctity of the land-owner, particularly if in residence on the site. Respect for the owner might involve limiting the number of parked cars, noise restrictions, or even access points.

The terms of the lease are critical to the continued use of the facility, and should be drawn up under the guidance of a lawyer who will also search title and discover any easements or encroachments existing on the parcel to be leased or on the remainder. "Squatting" on a parcel temporarily without a lease is to be discouraged, for not only will "squatting" invite ownership problems but also will place the observatory in jeopardy at the whim of the landowner. Due to the expense and effort of construction, siting without a lease or ownership is hazardous. The whole aspect of zoning and siting must be thoroughly investigated before any attempt at purchase or lease is undertaken. The dark sky requirement coupled with unobstructed horizon limits opportunities for sites, and a good Landscape Architect or Site Planner could be well worth contracting to select an optimum site on the land, since he/she will consider other environmental constraints that have not been considered.

Comparison of Roll-Off Roof and Domed Observatories

Criteria	Roll-off roof	Domed
Inside space/volume	More space for operator/visitors to move around in a rectangular format.	Less horizontal space but more vertical Space. Often confined to two or three observers.
Weatherproofing	Roof is easily made weather-proof on whole roof systems, although soffits and gables require extra protection.	Dome requires much attention to sealing the gore panels and slot door.
Sky view	Fully open to entire sky dome with roof retracted.	Slot has narrow aperture. Dome must be rotated to re-position the opening to various sky segments.
Thermal equilibrium	Telescope and accessories reach thermal equilibrium quickly with open roof.	Takes more time to reach thermal equilibrium. Mixing of ambient air outside the dome with internal air inside the dome can produce eddies at the slot opening.
Dew-up	Exposed fully to dew. Every surface inside cools to temp. lower than ambient air temp., and collects dew.	Instruments and surfaces inside rarely collect dew, Dew settles only on Dome outer surface.
Wind exposure	Very exposed—high winds will affect viewing.	Little wind effect other than wind buffeting some lateral shutters of a dome. (The roll over the dome type shutter prevents this).
Versatility	More versatile—often will accommodate several instruments on several piers.	Less volume horizontally, but more volume vertically, Limited to one pier—perhaps with dual instruments.
Snow and ice load	Significant amounts may require removal, as snow load increases stress on castors, track and gantry. If tracks are icy, some danger exists in rolling the roof back.	Snow slides off dome very easily. Track and castors are safe and protected inside the skirt. Weather has little effect on dome observatories if the dome is secured down tightly in wind storms.
Automation	Not easily accomplished and dangerous if not supervised.	Dome rotation is easily automated, although a motor driven slot door is more complex to engineer.
Site suitability	More difficult to site due to size and long rectangular footprint. Requires uniform stability in footings or flexure will occur between gantry and observatory proper.	Since structure is vertical it requires less space and less flexure occurs within a single round footing. Often the only choice in small, narrow lots.
Zoning and by-law conformity	May exceed by-law minimum in some states and towns, particularly if gantry is considered part of the structure.	Requires less footprint being round, and can be often exempt under the 100 ft^2 maximum if kept to 10 ft diameter.
Privacy	Less privacy from shoulders up.	Maximum privacy.
Cost	Usually much lower due to all-wood construction (track is the only exception—welding).	Higher fabrication cost due ribs, arches, and dome ring requiring bending in mill. Requirement for all-stainless fixtures, bolts, nuts, washers, etc., is costly.
Difficulty to build	Again, usually within the scope of a home handyman with the average inventory of tools. Only the track needs light welding by a contractor.	Requires much extra skill in cutting and fitting of gore panels to create the Dome. Riveting and tapping is required, along with metal-work on arches and ribs. Helio arc welding required on dome base angle.
Maintenance	Exterior track, and gantry need paint and stain only. If walls are finished in vinyl Board and Batten, they are maintenance-free.	Only the Dome requires paint, usually in 3 year intervals—a difficult task on a spherical surface. If walls are finished in vinyl Board and Batten they are maintenance free.

The Prestige of Ownership

Once you have completed the observatory, you will experience a personal pride in owning a structure that signals to others that you are indeed an astronomer. Several spin-offs will occur from such a structure. Local schools and institutions will likely be aware of the fact that you are more serious about astronomy as a hobby if they visit your observatory. Its presence will indicate your involvement and investment in the science of astronomy. It is likely that you will be asked to host viewing sessions for various organizations, including school groups, cubs etc. The press will also become interested since few people embark on such a project. In all, the observatory will increase and improve your public profile.

The observatory is a place where the universe becomes your private domain. It is a place where you can concentrate on the sky and eliminate the distractions of the world around you. It becomes an extension of your telescope, and in fact you will find that it becomes increasingly difficult to distinguish the importance of one over the other.

Note

1. Golub Leon, Pasachoff Jay M (2001) Nearest Star, The Exciting Science of Our Sun, Harvard University Press

Chapter 3

Options for Foundation Types and Laying Out the Observatory "Footprint"

Once you have studied all the previous requirements and are satisfied that you qualify with respect to all of the physical and legal parameters, you are ready to begin construction. Again, I cannot stress enough the importance of building an observatory that you will not outgrow. The observing limitations you experience today will not be what you can tolerate tomorrow. Larger aperture instruments will be within your economic grasp and larger CCD sensors will become common-place (the CCD sensor driving you to create a warm room for a computer). Both opportunities could make you wish you had constructed an observatory at least as large as the model promoted in this book.

Slab Footing Versus Sono Tube Footings

A complete poured cement slab under the observatory proper is the easiest to install, providing you have cement truck access, several wheelbarrow assistants, and some careful planning. It will save you the arduous task of excavating at least 16 holes for sono-tube footings and filling them with concrete. In addition, the alignment of the filled tubes is not easy and nor is the setting of the iron saddles in line to hold the floor joists. If you have an inaccessible site, then the sono-tube footings will have to be undertaken. However, most situations will permit a cement truck to fill wheel barrows which can be wheeled to the site. This technique usually becomes economical if one has several assistants to help, since the driver starts adding on a surcharge after an hour's time has passed. This means that you will have to prepare the route to the observatory to make sure the soils will support the continuous parade of heavy wheel barrows. This is usually achieved by laying construction planking down in endless fashion from the truck to the site.

A little bit of preparatory work at this point will save you grief when the cement truck arrives and the clock starts ticking (Figs. 3.1 and 3.2).

© Springer-Verlag New York 2016
J.S. Hicks, *Building a Roll-Off Roof or Dome Observatory*, The Patrick Moore
Practical Astronomy Series, DOI 10.1007/978-1-4939-3011-1_3

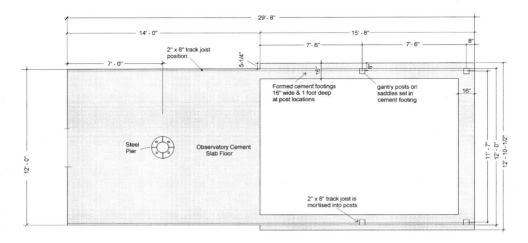

Fig. 3.1 Cement Slab Footing with cement footing extending out under the gantry section (Diagram by John Hicks)

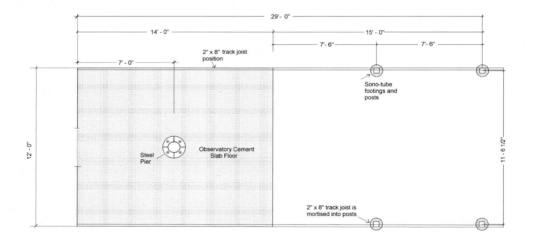

Fig. 3.2 Cement Slab Footing with Sono-tube piers under the gantry section (Diagram by John Hicks)

Preparing the Site and Orienting the Observatory

The first action you should take is to level off the site and smooth it to a good working surface, free of turf, roots, rocks etc. Once this is done, it helps to lay down a shallow depth of cement sand over the whole footing area to keep your boots from getting all gummed up with existing soil.

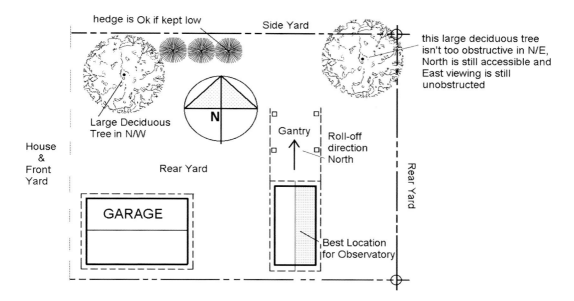

Fig. 3.3 Orientation of the observatory with roof rolled off northward is the best arrangement with no southern obstruction by the roof gable. Normally, except in very southern latitudes Polaris is still high enough above the roof gable to be imaged and allow for polar-aligning the telescope. In this set-up the garage, large trees and rear-yard set-back limit possible locations, but if the garage roof is low enough a fair western view is still possible. The Southern sky is completely accessible which is a priority (Diagram by John Hicks)

(In all likelihood, once you excavate for footings, wetter soil profiles will be encountered, and the wet, sticky material will get underfoot).

Following your requirement for a southern sky, the gantry section should be oriented to the north and observatory itself to the south. Of course, in the southern hemisphere the preferred orientation is reversed. This is not always possible, but usually the preferred arrangement. Rolled-off in the north position, the roof gives good weather protection from the cold north and north-westerly winds, whereas, if located with the observing room on the north side, it would receive the full force of these winds (Fig. 3.3).

Measuring and Squaring the Building Footprint

Measure off the length and width of the entire structure and site the entire footprint to comply with your local zoning restrictions. Establish the floor of the observatory itself first. In this design it will be 14 ft long × 12 ft wide. To assure squareness, measure both diagonals from corner-to-corner. They should be exactly equal if your measurements are true.

Repeat the length and width measurements, repositioning the stakes at the corners and adjusting all the stakes until the diagonal measure is satisfied. Once this is accomplished, locate the pier position at the center of the two diagonals with a string line and stake it firmly. The gantry posts should next be located measuring outwards from the corners of the observatory and the hypotenuse measured from the end posts until the ends are square with the observatory. Holding a string in line along the forms for the observatory slab, extend it out to the corner of each proposed end post, locate it and measure in to locate the middle posts (note that all these 5½″ × 5½″ posts are flush in line with the foundation edge which means their centers are offset inward by 2¾″). Make sure all the gantry stakes are firmly located at the center of the proposed posts, and drive a nail in at the exact center point in each stake (Fig. 3.4).

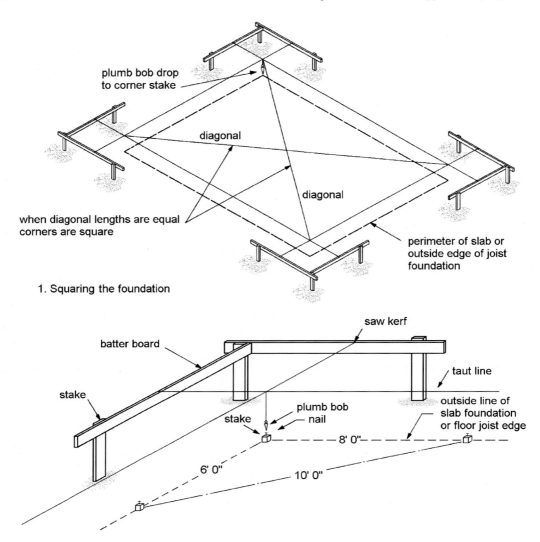

plumb bob drop
to corner stake

diagonal

diagonal

when diagonal lengths are equal
corners are square

perimeter of slab or
outside edge of joist
foundation

1. Squaring the foundation

saw kerf

batter board

stake

taut line

plumb bob
nail

stake

outside line of
slab foundation
or floor joist edge

8' 0"

6' 0"

10' 0"

2. Staking out the corners (by geometry)

METHOD OF STAKING OUT & SQUARING SLAB OR
SONO-TUBE FOUNDATION

Fig. 3.4 Foundation Layout—Measuring and Squaring the Observatory footprint (Diagram modified from Canadian Wood Frame Construction)[1]

Note

1. Canadian Wood Frame Construction, Central Mortgage and Housing Corporation

Chapter 4

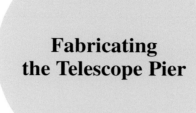

Fabricating
the Telescope Pier

Whether you choose the poured concrete slab or the sono-tube footings construction for your foundation, the pier must be located and poured first in position. At this time it is essential to decide which type of foundation you wish to install for two reasons:

(a) the electrical wiring to the pier should be installed in plastic pipe under or within the concrete slab if you plan to pour a complete cement floor.
(b) the height of the concrete portion of the pier (the pier footing) must be pre-calculated to a greater height above the ground in a sono-tube footing and wood joist style observatory (usually about 2 ft high).

The pier must go in first for another reason;—regardless of the footing type, the building wall height relates to the footing and the finished pier top. A poured concrete slab footing positions the finished floor about 8 in. above grade with the concrete pier footing top another 2 in. higher, whereas a sono-tube-plus joist style floor construction will place the pier footing top about 2 ft above grade. Also, in order to pour the concrete pier to a suitable height, *if you choose a totally poured concrete column*, the height of instrument calculation becomes essential now in order that the instrument doesn't get in the way of the rolling roof, opening or closing (Fig. 4.1).

In my own case, I used a column of concrete in a sono-tube to about one-half the height of the proposed pier above the floor, completing the rest with the same length of equal diameter thick-walled aluminum pipe. I machined heavy 1½″ thick aluminum flanges as caps over each end, recessed to fit into the pipe, and machine-bolted into the tapped flanges through the pipe walls. The flanges were pre-drilled to fit the bolts protruding from the concrete pier base on the bottom end (requiring a template to be made first in heavy card), and to match the bolt circle of the telescope mount on the top end. Also, circles were cut out of the pipe at positions convenient for access to the bolt threads inside in order to put the nuts on and tighten them with a wrench. When completed, I covered the whole length of pier inside with soft broadloom, sewing it all the way down the pier. This gave me "creature comfort" on cold clammy days when I had to straddle the pier or otherwise come in contact with it. The overall fabrication was difficult but worth the effort in appearance and function (Fig. 4.2).

© Springer-Verlag New York 2016
J.S. Hicks, *Building a Roll-Off Roof or Dome Observatory*, The Patrick Moore
Practical Astronomy Series, DOI 10.1007/978-1-4939-3011-1_4

CALCULATING THE HEIGHT OF THE PIER

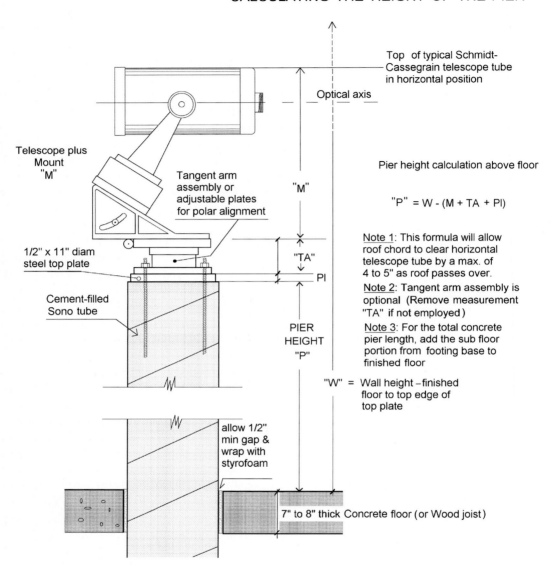

Fig. 4.1 Calculating the Pier height above the floor. A typical Schmidt-Cassegrain telescope on a fork mount is shown in horizontal position atop a full length sono-tube concrete pier. The cement-filled sono-tubed pier extends down to the footing. Calculating the height as shown will place the horizontal telescope tube under the bottom roof chord by about 4–5 in. (Diagram by John Hicks)

Fig. 4.2 Typical Aluminum or Steel Pier top showing end-plates: the top plate with bolts protruding to accept the telescope mount, the bottom plate with pre-drilled holes to fit bolts from cement footing underneath. Note the access-holes sawn into the tube required for tightening up the bolts from inside. Both top and bottom plates were drilled and tapped to take large threaded bolts. The access holes serve another purpose also; the tube can be filled with silica sand through these holes to dampen any vibration in the pier (Photo by John Hicks)

Combining a Concrete Pier Footing with Upper Metal Pier

The usual method of pier construction involves an oversize concrete footing with a much less diameter steel or aluminum extension to the telescope. In our design, the pier footing is 16 in. in diameter, preferably poured into a sono-tube form seated in a large excavation. Concrete could be poured simply into an excavation 16 in. in diameter, but it is always wiser to pour into a form for several reasons, the most important being that you might have to remove it some day. In frost zones, the excavation should go down 4 ft to assure that no frost heave can occur. If it is poured in a sono-tube footing, they are designed to allow for frost expanded soil to slide by the outside walls of the waxed tube and I strongly advise using one. I needed only a single 10″ diameter sono-tube in an excavation 4 ft deep with gravel bedding underneath and around it. I also only needed to support a light 100 mm diameter refractor or at most an 8″ Schmidt Cassegrain telescope. My soil profile is hard-pan clay, with no moisture. The soil resisted my digging like concrete. In other soil types with more sand or more water, you will need a wider footing with a flat base slab underneath.

Pouring the 24″ × 24″ × 8″ thick base slab is no easy task since the pier hole will have to be widened at its bottom, and some sort of a form made. Creating a perfect 24″ × 24″ footing is ideal, but simply pouring a flat concrete pad into an enlarged bottom will suffice.

Smooth out the top surface flat to seat the 16″ diameter sono-tube footing and let it dry. Insert the 16″ diameter sono-tube and fix it into position vertically with a carpenter's level. Add 3/4 crushed gravel around its sides up to soil surface level to hold it in position. Tamp the gravel into place and repeatedly check with a level. Fabricate a 5/8″ rebar cage by wiring eight strands of rebar together to form a basket about 12″ diameter. This should be in a column long enough to extend the length of the pier but also accommodate the bolts that will hold the steel upper pier portion. Leave enough room above the rebar cage for them. I used coat-hanger wire to fasten the cage together. Heavy duty pliers, a wire cutter, and hacksaw are needed to build the rebar cage.

Note that if you are using a full-length concrete pier leave clearance so the rebar cage does not interfere with the long mounting bolts that fasten the telescope mount to the pier.

Your finished concrete footing should terminate an inch or two above the finished concrete floor in a slab type observatory floor and about 2 ft above grade in a sono-tube and wood joist floor type. In both cases it should terminate above the floor so that you can get at the bolts securing the rest of the metal pier. When pouring the concrete into the sono-tube form, continuously tamp the wet concrete around the rebar cage with a long stick to eliminate air pockets and voids in the cement. In preparation for pouring the pier base, you must also prepare a pier base template in 3/4″ plywood drilled accurately to hold the anchor bolts in place which will hold the metal top section of the pier. This means you will have to design the metal pier portion at least on paper beforehand, and drill a bolt circle in the plywood template to hold the anchor bolts matching the design of your metal pier base. It is safer to proceed this way, rather than attempting to fabricate the entire metal pier portion first, and find the concrete pier footing has a shortfall. The metal pier portion can always be made longer once you measure the height of the finished wall, which will enable you to more accurately locate your telescope height just where you want it to be (Figs. 4.3, 4.4, and 4.5).

Place nuts on both sides of the plywood template, securing the threaded rods at the correct length to anchor into both the concrete pier below and the future metal pier base above. When the pour reaches the top of the sono-tube footing, plunge the threaded-rod-bolt assembly into the concrete, center and level it. Trim off any excess concrete that oozes out from under the template.

This is a very critical step, so perform it carefully. Before pouring concrete, make sure that there is sufficient room between the top of the sono-tube and the inserted rebar cage. Try it with your plywood template before you pour the concrete. Some builders prefer at this point to leave the nuts on the underside of the template also, to act as good "seats" for the metal pier section, the nuts staying flush in the concrete once the plywood template is removed (Figs. 4.6 and 4.7).

Full Length Concrete Pier

In pouring a full-length concrete pier, you will have to pre-calculate the height your telescope and mount require with the telescope in a horizontal position. With the top plate of the wall (holding the castor track) as maximum height, the sono-tube form length then equals the below-floor-length to footing, plus above floor calculation. Refer to Fig. 4.1 to review the required calculation.

Be aware that when pouring the a high concrete column you will have to brace the sono-tube with stakes and boards to the ground. A pier 5 ft high will start to lean under its own weight and the sono-tube can rupture or split. If a full length concrete pier is poured in place with no metal extension, you will also have to make a pier cap with a bolt circle that matches your mount for the top. In this case, you should use a steel or aluminum cap about 3/4″–1″ thick. Trace out the base of your telescope mount complete with its mounting holes on paper card stock precisely. Transfer the paper card stock

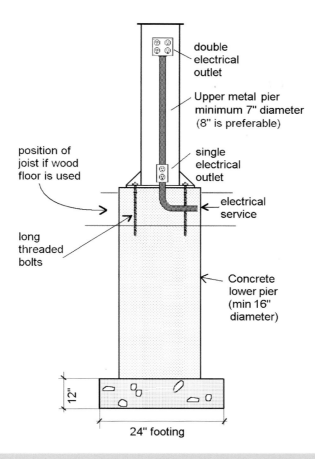

double
electrical
outlet

Upper metal pier
minimum 7" diameter
(8" is preferable)

position of
joist if wood
floor is used

single
electrical
outlet

electrical
service

long
threaded
bolts

Concrete
lower pier
(min 16"
diameter)

12"

24" footing

Fig. 4.3 Detail of finished metal pier top on its concrete foundation showing electrical conduit, electrical outlets, and bolts anchoring the metal top to the concrete footing underneath. Normally the metal pier top should begin about 2 in. above your observatory floor. The faint image of a joist is shown to illustrate where floor joists would sit in a sono-tube-wood-floor foundation. From this you can visualize that the steel pier base rests about 2″ above the finished floor surface in a concrete slab or wood floor foundation (Diagram by John Hicks)

trace to the metal cap and drill the holes for the mounting bolts that will extend well into the concrete mix when you pour the concrete into the pier. (My bolts were 18 inches into the concrete pour and extended 1 inch above the metal pier cap). Secure with nuts above and below (the bottom nuts can stay in the concrete with the cap, but the top nuts are removed and replaced on top of your telescope mount to secure it) (Fig. 4.8).

If you cannot find a blacksmith or machinist to fabricate a cap for you, check out your local scrap yard. Boiler punch-outs are perfect for this purpose and most are quite thick.

If you find one close to the pier diameter take it home, have it machined smooth to your exact concrete pier diameter, and drill through it to accept the mounting bolts. If your telescope mount has projecting bolts then you will have to drill and tap the pier cap for them. I was able to get my pier cap machined by a local machinist and drilled the holes myself on a drill press.

This cap is pushed into the top of the finished cement pour and leveled. Check every 15 min or so as the concrete dries to adjust for level. If you are using a full length concrete-filled pier, you will want several electrical box outlets on the side of the concrete column. Before the concrete pour dries, drill through the sono tube using the junction box bolt holes as a template. Insert Stainless

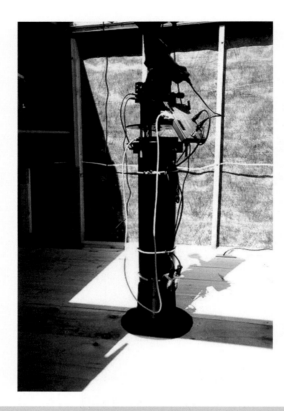

Fig. 4.4 A well-planned steel pier resting on a concrete footing underneath the floor. Notice the control pad shelf just under the mount. With the increasing complexity of computerized controls etc., the need for multiple electrical sockets becomes apparent (Courtesy of Terry Ussher)

Steel mounting bolts in the box and push them through the sono-tube walls into the wet concrete. Leave these in while the concrete dries. (I prefer to use long S/S round headed 1/4 × 20 bolts, cutting off the heads, and fastening stainless steel nuts to eventually tighten down the box, allowing removal at some later date) (Fig. 4.9).

The pier must have no connection to the concrete floor or the wood joist floor to avoid vibrations produced by people walking around in the observatory. If a poured concrete floor is used, wrap a 1″ thick collar of Styrofoam around the pier at least a foot deep at this stage before pouring any concrete into the forms.

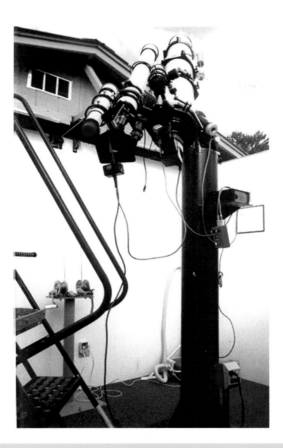

Fig. 4.5 The stronger the steel pier the better. This pier carries multiple telescopes which add up to a lot of weight (From the collection of John Hicks)

TYPICAL METAL UPPER PIER DESIGN

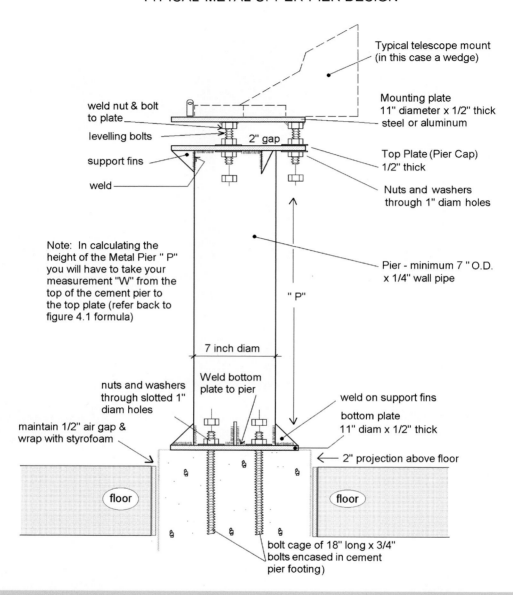

Fig. 4.6 Detail showing construction of typical metal pier. Note cement-filled oversize sono tube underneath and the level of finished floor (Diagram adapted from AstroImage Database)[1]

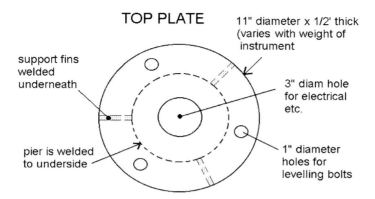

TOP PLATE

11" diameter x 1/2' thick
(varies with weight of
instrument

support fins
welded
underneath

3" diam hole
for electrical
etc.

pier is welded
to underside

1" diameter
holes for
levelling bolts

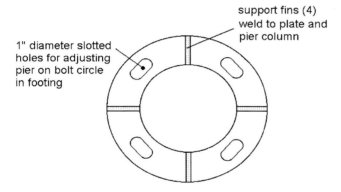

BOTTOM PLATE

support fins (4)
weld to plate and
pier column

1" diameter slotted
holes for adjusting
pier on bolt circle
in footing

Fig. 4.7 Detail showing construction of typical metal pier caps (Diagram by John Hicks)

Fig. 4.8 Typical all-cement pier, enclosed in a Sono-tube form, with bolt heads protruding from the steel-plate top cap, ready to accept the telescope mount (Diagram by John Hicks)

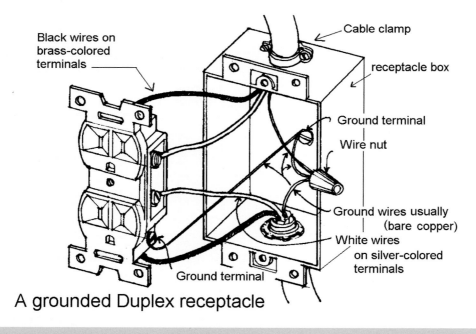

Black wires on
brass-colored
terminals

Cable clamp

receptacle box

Ground terminal

Wire nut

Ground wires usually
(bare copper)

White wires
on silver-colored
terminals

Ground terminal

A grounded Duplex receptacle

Fig. 4.9 Wiring a Grounded Duplex Receptacle (Diagram adapted from Home Renovation)[2]

Notes

1. David's astronomy Pages, AstroImage Database, Pier to Mount Engineering Drawing. Web address:http://www.richweb.fg.co.uk/astro/imagelib/PierDrawing.html
2. Ching, Francis D.K & Miller, Dale E. (1983) Home Renovation, Van Nostram Reinhold Company Inc.

Chapter 5

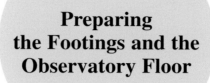

Preparing
the Footings and the
Observatory Floor

The Concrete Slab Floor

Preparing the Forms for the Observatory Floor

Referring to the following plans, you will notice that the thickness of the concrete slab increases under the walls around the perimeter of the observatory, and under each of the posts supporting the gantry section (Figs. 5.1 and 5.2).

The normal floor thickness should be about 7½″ thick throughout increasing to 12″ under the walls and posts. Prior to pouring any concrete, forms must be firmly in place to retain the concrete and create a finished, clean edge about the perimeter. Since the concrete is heavy, it will attempt to push out any substandard forms you use, so use heavy boards and substantial stakes to hold them in place. I recommend using 2″×8″ boards nailed to 2″×2″ stakes. (The concrete can be cleaned off and the boards used later in framing).

Carefully line the boards on edge along a string line stretched between the corner stakes, nailing them carefully to the stakes. Drive the stakes in every 2 ft securely all around the floor perimeter. Excavate shallow 4″–6″ trenches at least a foot wide under the wall sections. This will create a footing ~12″ deep×12″ wide to support the walls. Make sure you align the tops of the forms so that they are placed exactly where the top of the finished floor will be (use a level all around the foundation) as they will serve as the rails for running the "screed board" to level and finish off the floor.

Preparing the Forms for the Gantry Section

The gantry footing is offset from the observatory walls to create the proper positioning of the track and frame over the gantry posts (they must be right under the center-line of the track to carry the load of the roll-off roof). Refer back to Fig. 3.1. You can see the offset in the gantry footing better in the end view (Fig. 5.3).

© Springer-Verlag New York 2016

J.S. Hicks, *Building a Roll-Off Roof or Dome Observatory*, The Patrick Moore
Practical Astronomy Series, DOI 10.1007/978-1-4939-3011-1_5

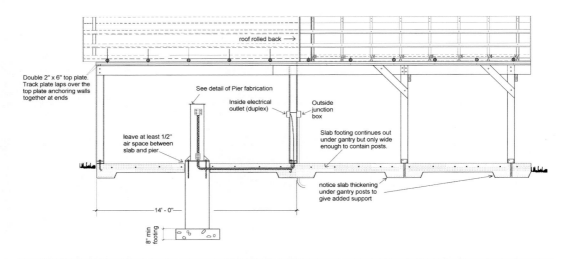

Fig. 5.1 Elevation view of observatory cement slab floor and gantry slab footing (Diagram by John Hicks)

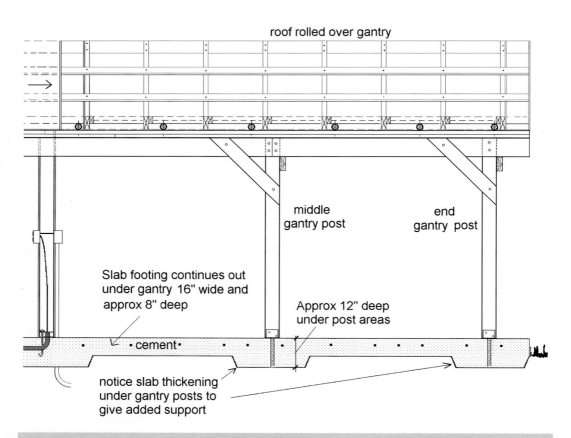

Fig. 5.2 Enlarged view of concrete footing under gantry (Diagram by John Hicks)

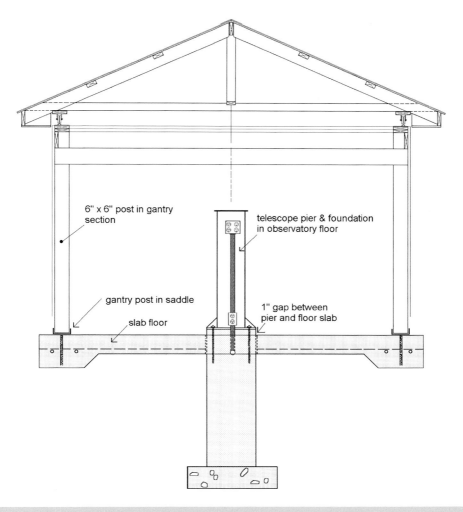

6" x 6" post in gantry section

telescope pier & foundation in observatory floor

gantry post in saddle

1" gap between pier and floor slab

slab floor

Fig. 5.3 End elevation of gantry section (Diagram by John Hicks)

The entire gantry footing should be about 16″ wide and 7-½″–8″ deep, increasing to a foot deep under the posts, in order to provide enough mass for concealing the rebar reinforcing and support for the roof once it is overhead (Fig. 5.2). The need for a completely poured floor extending all the way from observatory section out under the gantry is up to the individual. It requires a great deal more concrete, lots of effort, and is unneccessary. Finally, insert lengths of 5/8″ rebar in double rows in the concrete at half-filled stage around the perimeter to prevent cracking and displacement.

Line out the forms for the gantry footing same as you did for the observatory floor, following the measurements on Fig. 3.1 showing the complete concrete footing. The forms that you used to line out the floor of the observatory will act as guides for locating the outside face of the gantry posts. The track sits on the center line of the beams mortised into the face of these posts. Measure out the distances shown from the observatory slab, and follow the measurements on the plan to locate the post centers and the edges of the footings. Notice that they are offset from the observatory floor edge. The offset places the post in the center of the concrete providing equal bearing all around it. Leaving the center of the gantry section open will allow you to fill this area with a lockstone patio later on, providing a neat concrete edge all around, and an attractive outdoor sitting area.

Electrical Service

In the case of a full concrete floor, before concrete is poured, you must prepare your electrical service lines and computer/telephone feed wires in two adjacent PVC pipes (1½″–2″ I.D. plumbing pipe). These must be long enough to reach from the pier to the wall where you have chosen to locate an electrical outlet. The PVC pipes at the wall will come up through holes drilled in the sole plates (the 2″×6″ plate under the studs) through the hollow wall framing to a receptacle. It is best to locate a sealed junction box on the outside wall at this point also, so the connecting wires just go through a short section of wall to connect to the underground wire source from your house (Fig. 5.4).

Number 14 G wire is suitable for the inside portion of the observatory, under the floor in PVC pipes and up the walls, but 8 G wire should be used for distances to the house. To get the wires through the PVC pipes, feed light fishing line with a sinker attached first, pulling the wires with it. Both ends of each PVC pipe are then fitted with rounded fittings such that the wires will go up the pier on one end, and up the wall cavity on the other end. Measure the PVC pipes carefully, trying them in the footings so you know they will reach from pier to wall. Make sure the wall exit end does so in a mid-stud position by measuring in from a corner. When satisfied, leave about a yard or so of wire projecting at each end and cut off the balance. Wire or rope the ends of the rounded fittings at the pier such that the

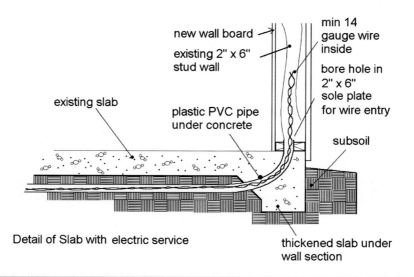

Detail of Slab with electric service

Fig. 5.4 Detail of electrical service under slab and through wall sole plate (Diagram by John Hicks)

fittings just emerge about an inch or so above the level of the finished floor, and stake the fittings upright at the wall plate position with the same amount projecting. Coil up the wire at the wall end, wrapping it in plastic. Loop the wire around the pier at the pier end, tying it with a rope to keep it free of the concrete. Make sure both pipes are flat on the ground so that they will remain at the base of the concrete floor, then cover the entire ground with construction grade 6 mil plastic sheeting.

Pouring the Concrete Floor

Make sure to provide a ramp up onto the form boards to get the wheelbarrows into the center area of the observatory, and nail it to the forms securely well in advance of the cement truck arriving. Also, get the pathway from the truck to the observatory lined out in rough planks, and as level as you can make it. The wheelbarrows full of concrete are so heavy and cumbersome they will sink into your lawn and mire.

Starting at the center area of the floor around the pier, dump the first loads of concrete, working outward in all directions to the outside walls. Keep the concrete pour from touching the pier by wrapping the pier base with ½″ styrofoam sheet wrapped around the pier base.

As dumping proceeds, rake the concrete level out over the floor area. When about ½ full, drop in the sheets of 6″×6″ #8 welded wire mesh covering most of the floor area. Avoid placing it near the location of the walls because that is where the anchor bolts (j-bolts) will have to be set. Instead of the wire mesh, you can use 5/8 in. diameter rebar, cut to fit in a radial fashion outward around the pier. It should extend, as above, just short of the anchor bolt areas. Be aware that neither of the reinforcing materials are bent upward, and lie flat on the bed of concrete, or they will protrude from the finished floor. When the concrete level reaches the batter board height (about 7½″), smooth out the surface with a screed board to flatten out the floor. This is accomplished with a long board stretching right across the width of the floor, resting on its edge upon the tops of the batter boards.

Finish all the floor area as far out as your arm can extend, with a cement trowel, smoothing out the surface flat to make a good seat for the wall sole plates.

Positioning the J-bolts

The next step, setting the j-bolts in position quickly, before concrete dries, is fairly critical. The j-bolts holding the building to the slab, and the saddles (or rods) holding the gantry posts must now be pushed into the wet concrete at the appropriate locations.

For the observatory walls with 5½″ wide sole plates, the j-bolts should be placed inside the concrete edge 2½″–3″ about every yard apart. Measure in to locate each j-bolt and push it into the wet concrete with about 2½″ thread exposed above it. Keep the concrete off the bolt threads. Do not locate a j-bolt where the door opening is to be located, as the door threshold will be placed directly on the slab (Fig. 5.5).

Positioning the Gantry Posts

The gantry posts can be set in steel saddles for 6″×6″ posts, or they can be pre-drilled to fit over thick rods or re-bar set in the cement footing. If using rods or rebar leave about 4″–6″ exposed above the concrete. I prefer steel saddles because they don't weaken the post bases and a post can be removed

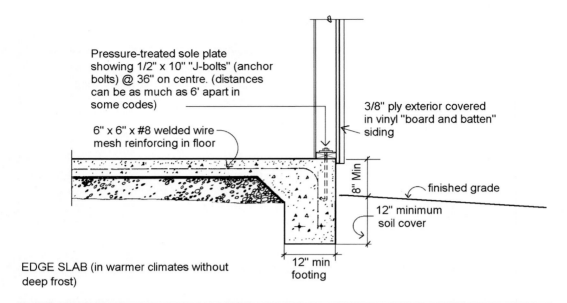

Pressure-treated sole plate
showing 1/2" x 10" "J-bolts" (anchor
bolts) @ 36" on centre. (distances
can be as much as 6' apart in
some codes)

6" x 6" x #8 welded wire
mesh reinforcing in floor

3/8" ply exterior covered
in vinyl "board and batten"
siding

8" Min

finished grade

12" minimum
soil cover

EDGE SLAB (in warmer climates without
deep frost)

12" min
footing

Fig. 5.5 Detail of floor slab and sole plate with "j-bolts" (Diagram adapted from Home Renovation, for further Reading see note (2) Chap. 4)

and re-set easily in the saddle. Pull out the stakes you used to locate the exact position of the posts, and quickly insert the steel saddle or rod, pushing them deep into the concrete footing. In using steel saddles, you will have to align them relative to each other, accomplished easily with a section of $2'' \times 6''$ board, placed flat in them. The saddle should sit on the finished concrete surface. If hot-dip galvanized (preferred) they can rest right on the concrete surface which lends greater support (less torque on the rod supporting it). A normal ferrous saddle could be used but must be raised to prevent corrosion at the concrete face.

Check the concrete before drying, maintaining that the j-bolts, saddles or rods remain perpendicular and in-line. Toward evening, lightly sprinkle the concrete surface with a fine mist of water from a garden hose so that the mixture does not dry too fast and crack. When concrete is hard, strip off the forms and remove the stakes in preparation for framing the walls.

The Sono-Tube Footings Floor

This footing arrangement requires a total of 16 sono-tube footings (12 for the observatory floor and 4 for the gantry posts). With the $2'' \times 10''$ heavy floor joists this footing arrangement will provide all the support you will ever need, even with a crowd in the observatory. If one were to use a smaller size joist such as a $2'' \times 8''$, additional footings would be required as shown in the second diagram with 16 footings under the observatory floor (Figs. 5.6, 5.7, 5.8, 5.9, 5.10, 5.11, and 5.12).

The joists forming the sub-floor of the observatory are held in place over the footings with iron saddles set in concrete. These will allow for some adjustment, shimming, etc., in case you make a mistake in squaring the foundation. Construction measurements should always be taken from corner "batter boards" staked at the outside corners. These are offset from the actual corner footings by 2 ft to allow space for digging holes for the sono-tube footings. If you add exactly 2 ft on both ends and

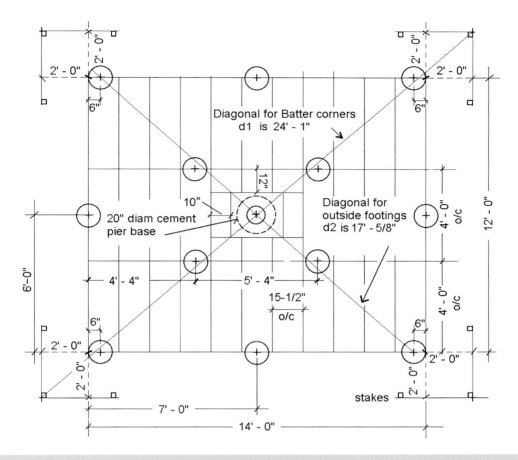

Fig. 5.6 Plan view of 12 sono-tube footing arrangement showing joist pattern (Diagram by John Hicks)

sides of the staked floor dimensions you will end up with a rectangle 18 ft×16 ft measured from the "batter board" corners. Set the batterboards to describe an 18 ft×16 ft rectangle. Measure the diagonals, corner to corner of the "batter boards" to make sure they are equal.

The sono-tube corner footings are measured in 2 ft from batter boards on the sides but 2 ft 6 in. from batter boards on the ends (the 12 ft dimension).

Referring to the diagrams of sono-tube footings, you will notice that the end footings are an additional 6 in. in from the batter boards—to avoid clashing the joist saddles with the floor joist junctions at the corners. Drive in stakes at the four corner footing locations. It will soon be apparent why we use offsets to the actual footing locations: because once the holes are dug, one loses any idea where the exact center of the footing is. In order to square the footings, again the diagonals should be equal. Measure from corner to corner diagonally.

All the floor joist saddles are arranged so as not to clash with cross joists. This also allows for easier nailing at the corners.

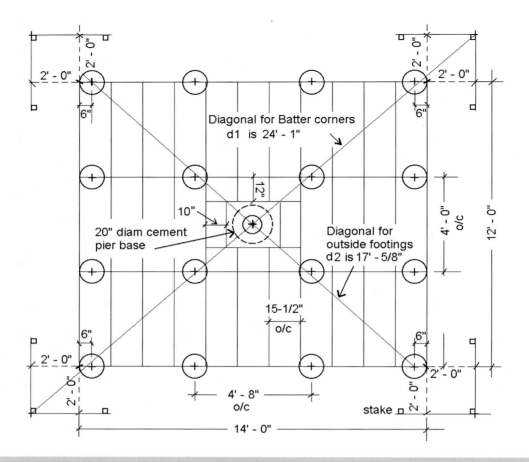

Fig. 5.7 Elevation of 16 sono-tube footing arrangement showing joist pattern (Diagram by John Hicks)

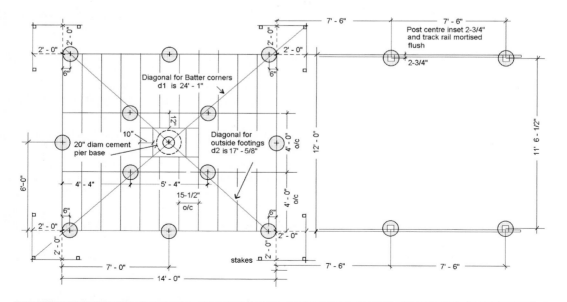

Fig. 5.8 Plan view of complete sono-tube footing arrangement including gantry showing 12-footing observatory foundation (Diagram by John Hicks)

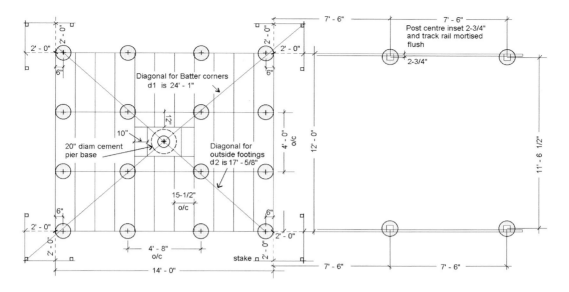

Fig. 5.9 Plan view of complete sono-tube footing arrangement including gantry showing 16-footing observatory foundation (Diagram by John Hicks)

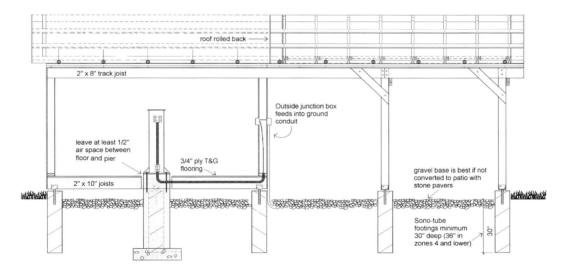

Fig. 5.10 Elevation of sono-tube footings arrangement—observatory and gantry (Diagram by John Hicks)

Observatory Floor Framing

The wood floor framing is constructed of even-spaced joists covered in a layer of 5/8″ tongue and groove plywood sub-flooring. Joists are 2″ × 10″ to hold the weight of a group in a concentrated viewing session to guarantee little or no deflection in the floor.

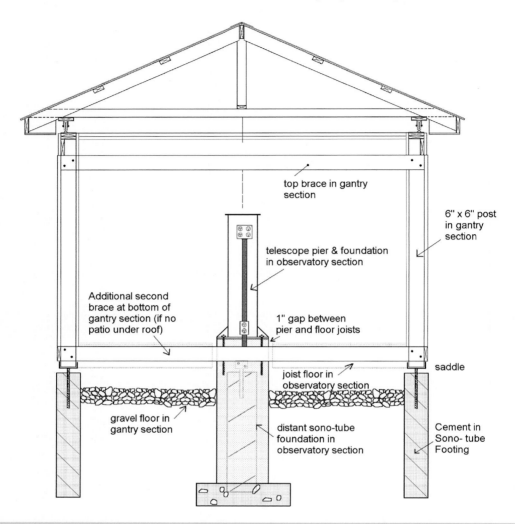

top brace in gantry
section

6" x 6" post
in gantry
section

telescope pier & foundation
in observatory section

Additional second
brace at bottom of
gantry section (if no
patio under roof)

1" gap between
pier and floor joists

saddle

joist floor in
observatory section

gravel floor in
gantry section

distant sono-tube
foundation in
observatory section

Cement in
Sono- tube
Footing

Fig. 5.11 End view of gantry with sono-tube footings showing pier in distance (Diagram by John Hicks)

With 16″ spacing, these will easily support a live load of 40 lb/ft^2.

Normally 2″ × 8″ joists would suffice in an application as small as this, but a rigid floor is essential for minimum movement. Use 14 ft 2″ × 10″ beams for the outer framing on both sides with no joints. Position and nail these first in the saddles, verifying the squareness by measuring the diagonals from corner to corner. Use 3–4 in. spiral ardox nails to nail the ends together and to end-nail all the joists (if you do not use joist hangers).

Position the 2″ × 10″ joists on 16 in. centers and nail with joist hangers (preferred method—stronger and meets newer building codes).

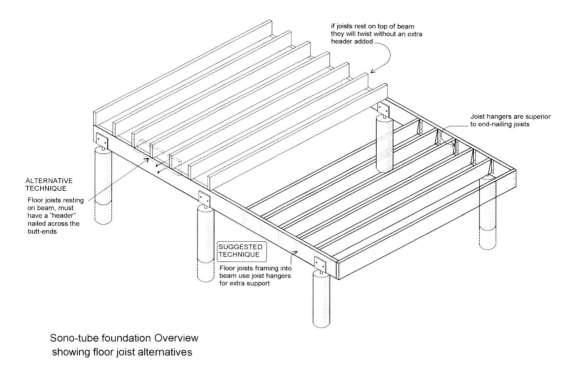

if joists rest on top of beam
they will twist without an extra
header added

Joist hangers are superior
to end-nailing joists

ALTERNATIVE
TECHNIQUE

Floor joists resting
on beam, must
have a "header"
nailed across the
butt-ends

SUGGESTED
TECHNIQUE

Floor joists framing into
beam use joist hangers
for extra support

Sono-tube foundation Overview
showing floor joist alternatives

Fig. 5.12 Overview of typical sono-tube footings with choice of joist placement (Diagram adapted from Home Renovation—see (2) in Further Reading of Chap. 4)

Combining a Poured Cement Observatory Floor with Sono-Tube Footings Under the Gantry

In some locations with no threat of frost and differential soil heaving, a combination of footings could be used to simplify construction. Using a poured concrete slab for the floor and four sono-tube footings for the gantry posts will make the footing task easier, eliminating forms for the gantry section. Such an arrangement introduces a differential foundation support which could allow flexing at the gantry/observatory junction. This is somewhat catastrophic because the track would bend at this point, preventing the roof from rolling off easily or perhaps entirely. In perfect conditions of soil and moisture, and no frost, the combination will work well. Consider this very carefully before you attempt it. Personally, I do not recommend it (Figs. 5.13 and 5.14).

Converting the Gantry Section into an Outdoor Patio

If you follow the route of pouring a solid concrete floor for both observatory and gantry perimeter, a stone paver patio can be installed interior to the gantry strip footings. This is the ideal finishing touch to an otherwise "just functional" observatory. When the roof is rolled off for solar observing etc., you can sit in the shade of the roof rolled off over the gantry. Even when the roof is closed, at the correct daylight hour, a sunny or shady nook is still created. The concrete margins, clean and sharp, add an

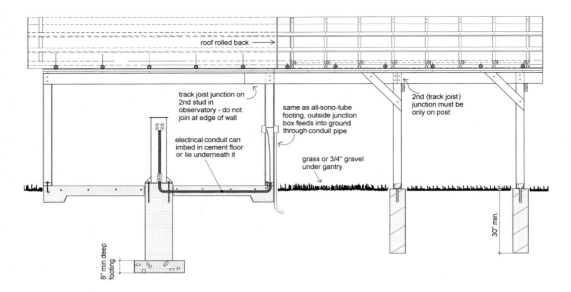

Fig. 5.13 Side elevation of combined footings—slab under observatory and sono-tubes under gantry (Diagram by John Hicks)

ideal border to the stone paver patio, creating a truly professional touch. If you were to use the sono-tube footings, it can still be done, but with much less effect, since the sono-tubes will interfere with the installation of a "clean" stone paver margin. You will have to cut the stones to fit and make some sort of border in wood or concrete to hold them in. You save all this extra work with a purely poured footing all around (Fig. 5.15).

Installing the Floor

The finished sub floor of the sono-tube footings model should be covered in a layer of outdoor 5/8″ tongue and groove plywood. This should preferably be screwed down which allows tightening later as the plywood dries out. Make sure to use at least coated screws or preferably hot-dipped galvanized type deck screws.

Cut the ply to fit around the pier base leaving at least ½″ gap all around. Do not allow the ply to touch the pier at any point. I wrapped a ½″ thick sheet of Styrofoam around my pier to seal the floor from drafts underneath. The styrofoam is cellular enough that it will not transmit vibration even if touching the pier. It serves a double purpose also a vibration dampening and dust barrier as the draft created through the track gap will pull air up through this opening depending on wind velocity. Note that the sole plate is nailed through the ply floor (Fig. 5.16).

The finished concrete floor of the poured foundation model can be insulated and covered in ply to create a "warmer" less damp floor to stand on. Installing 2″×4″ or 2″×2″ battens across the concrete floor in a lattice formation (at about 2 ft²), cover as above with 5/8″ tongue and groove plywood. Filling in the spaces with Dow SM board or similar will aid in your comfort on the cold winter nights (Fig. 5.17).

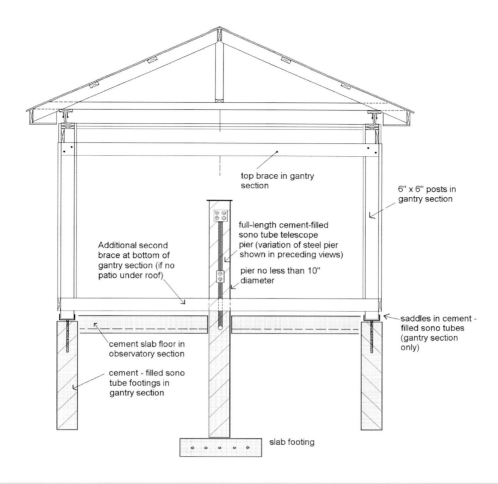

top brace in gantry section

6" x 6" posts in gantry section

full-length cement-filled sono tube telescope pier (variation of steel pier shown in preceding views)

pier no less than 10" diameter

Additional second brace at bottom of gantry section (if no patio under roof)

saddles in cement - filled sono tubes (gantry section only)

cement slab floor in observatory section

cement - filled sono tube footings in gantry section

slab footing

Fig. 5.14 End elevation of combined footings—slab under observatory and sono-tubes under gantry (Diagram by John Hicks)

Fig. 5.15 Photo of "Davis Memorial Observatory" with roof over patio (From the collection of John Hicks)

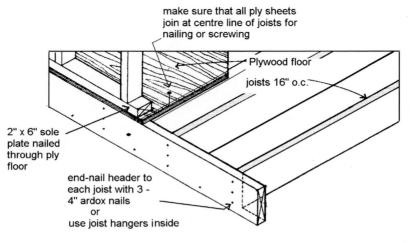

make sure that all ply sheets
join at centre line of joists for
nailing or screwing

Plywood floor

joists 16" o.c.

2" x 6" sole
plate nailed
through ply
floor

end-nail header to
each joist with 3 -
4" ardox nails
or
use joist hangers inside

JOIST and PLY FLOOR DETAIL

Fig. 5.16 Detail of floor framing (Diagram adapted from Home Renovation—see (2) in Further Reading of Chap. 4)

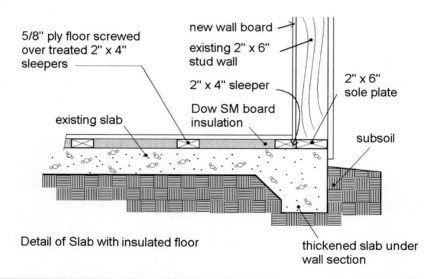

5/8" ply floor screwed
over treated 2" x 4"
sleepers

new wall board

existing 2" x 6"
stud wall

2" x 4" sleeper

Dow SM board
insulation

existing slab

2" x 6"
sole plate

subsoil

Detail of Slab with insulated floor

thickened slab under
wall section

Fig. 5.17 Detail of concrete slab with insulated floor (Diagram adapted from Home Renovation—see (2) in Further Reading of Chap. 4)

Chapter 6

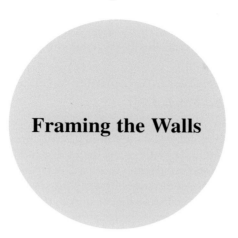

Framing the Walls

Constructing and Erecting the Walls

The four walls are best constructed on the ground, later to be raised up into position either on the joisted floor or poured cement floor with its j-bolts. Studs should be "end-nailed" through the sole plate (the bottom $2'' \times 6''$ plate) at a standard $16''$ apart on center. This method is much more satisfactory than installing the sole plate first and "toe-nailing" the studs into it while standing. Before installing the studs, depending upon the foundation type, it is essential to complete the following preparatory tasks:

(a) Locate the position of the j-bolts on the sole plate (in the case of a poured slab).
(b) Mortise the upper section of each of the side wall studs to accommodate the $2'' \times 8''$ track joist.

Locating the Position of the J-Bolts on the Sole Plate

Carefully position the sole plate on the threaded bolt ends projecting from the concrete slab, aligning it with the ends and flush with the sides of the slab. If you set the j-bolts at the same height in the concrete, you can simply hit the sole plate with a hammer impressing the bolt ends into the wood. If there is some difference in height of the j-bolts, carefully mark their position with a magic marker. Drill the holes for the bolts slightly larger than the thread size so the bolts won't bind in the holes. No matter how carefully you do this it won't be easy to lift up the whole wall and place it over the bolts without some binding—it pays to be very accurate locating the bolt ends. Make sure to try the drilled sole plate over the j-bolts **before** you begin nailing the studs to it, and mark the respective sole plate side (East or West). This may seem to be over-cautious, but when you are lifting the weight of the whole side wall, you will be thankful of the knowledge that it has been pre-fitted successfully. Don't forget to pre-drill the holes for the electrical conduit protruding from the poured concrete foundation also (mark their location at the same time you mark the j-bolt locations in the sole plate).

© Springer-Verlag New York 2016

J.S. Hicks, *Building a Roll-Off Roof or Dome Observatory*, The Patrick Moore
Practical Astronomy Series, DOI 10.1007/978-1-4939-3011-1_6

Wire from these passes up through the interior of the wall to a receptacle. It is a good idea to brace the studs horizontally with small sections of 2″×6″ placed midway between the sole plate and the top plate to add extra strength to the walls—particularly when you raise them up into position. These will also provide an extra support across the walls when you apply interior sheathing. They will have to be staggered up and down on the studs in order that you can butt-nail them.

Cutting the Mortise for the Track Joist

To support the track and rolling roof the side walls need a 2″×8″ track joist installed flush in a mortise cut at the top of the studs. It is far better to cut the mortise for the track joists now before the studs are attached. Cut the mortise deep enough that the 2″×8″ joist will be flush with the outside surface of the 2″×6″ stud, and also long enough that the joist sits perfectly flush with the top of the stud. Cut the mortise in all the studs before nailing them onto the sole plate.

Do not extend the track joist all the way to the corner nearest the gantry, leave an open 16″ gap in order that the next section of joist over the gantry fits inside the observatory wall providing an internal support for the continuing outside joist. This is necessary because beginning a new joist section outside the observatory walls introduces a weak junction and a possible flexure point. This means that the track joist mortised into the studs will fall short of the wall length required by 16″. It also requires that the corner stud(s) be cut shorter to accommodate it.

To assist in tying the studs all together, you can nail on the 2″×6″ top plate, end-nailing it to the top of the studs thereby securing the whole wall as a rigid unit (see Figs. 6.1 and 6.2).

Raising the Wall Sections and Tying the Frame Together

Raising the assembled wall sections is typically a two or three person job, so enlisting friends or astronomy associates is essential. Once a wall is lifted into place it should be temporarily secured with 2″×4″ braces, holding it in position until an adjoining wall is also lifted into place. Square the walls plumb with a carpenter's level and shim if necessary before nailing them together at the corners. Lastly secure the walls to the foundation using nuts on the j-bolts in a cement slab floor, or with nails in a joist floor (Fig. 6.3).

Looking at the detail showing the "outside corner assembly," notice how the 2″×4″ filler blocks are placed—the corner stud is followed by filler blocks sandwiched between it and the next 2″×6″ stud nailed together. The innermost 2″×6″ serves as a nailing stud for the interior wall sheathing. This arrangement may look over-structured, but it is a practical assembly method assuring both a rigid support and a means of fastening the exterior and interior sheathing at the corners (Fig. 6.4).

The top plate is a double top plate for strength in carrying the roof load. It consists of a lower 2″×6″ plate capping the wall studs and interior track joists, with another 2″×6″ track plate nailed on top. The 2″×6″ top plate nailed directly to the studs caps the wall section completely, sealing off the top of the observatory walls (prevents insects, such as wasps etc., from entering down into the wall sections).

Make sure to lap the second 2″×6″over all joints in the lower 2″×6″ plate, particularly at the wall junctions where the lap is essential to help hold the walls together.

In the next stage "building and aligning the gantry section" the track joists outside are capped in a double 2″×6″ aligned precisely with the inside 2″×6″ upper plates. The steel track base is lag-bolted directly down onto this, covering the top of the 2″×6″ surface completely. Since the track joists penetrate into the observatory about 16″ the gantry/observatory connection is well supported from any flexure (Fig. 6.5).

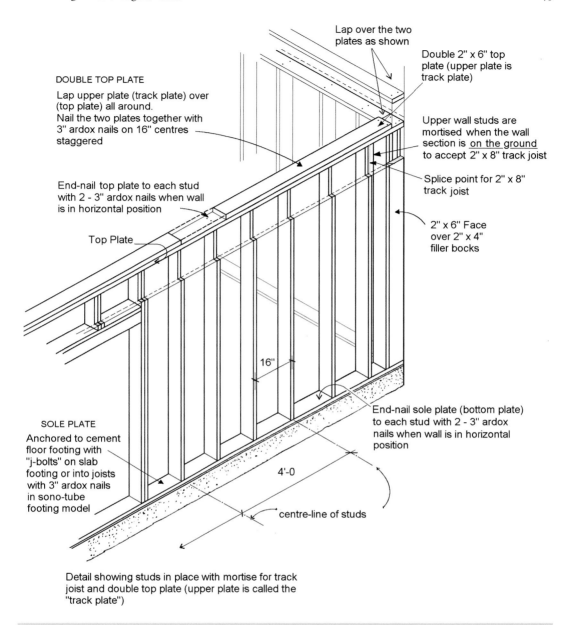

Lap over the two plates as shown

Double 2" x 6" top plate (upper plate is track plate)

DOUBLE TOP PLATE

Lap upper plate (track plate) over (top plate) all around.
Nail the two plates together with 3" ardox nails on 16" centres staggered

Upper wall studs are mortised when the wall section is on the ground to accept 2" x 8" track joist

Splice point for 2" x 8" track joist

End-nail top plate to each stud with 2 - 3" ardox nails when wall is in horizontal position

2" x 6" Face over 2" x 4" filler bocks

Top Plate

16"

SOLE PLATE

Anchored to cement floor footing with "j-bolts" on slab footing or into joists with 3" ardox nails in sono-tube footing model

End-nail sole plate (bottom plate) to each stud with 2 - 3" ardox nails when wall is in horizontal position

4'-0

centre-line of studs

Detail showing studs in place with mortise for track joist and double top plate (upper plate is called the "track plate")

Fig. 6.1 Detail of wall showing studs, track joist, sole plate, mortise, and top plate(s) (Diagram adapted from Home Renovation—see (2) in Further Reading of Chap. 4)

Nail the $2'' \times 6''$ plate to the track joist and wall stud ends with $4''$ ardox nails. Add the second $2'' \times 6''$ upper plate with $3\frac{1}{2}''$ ardox nails at ~$16''$ on center. When nailing on the double $2'' \times 6''$ plates make sure they are flush with the outside face of the track joist.

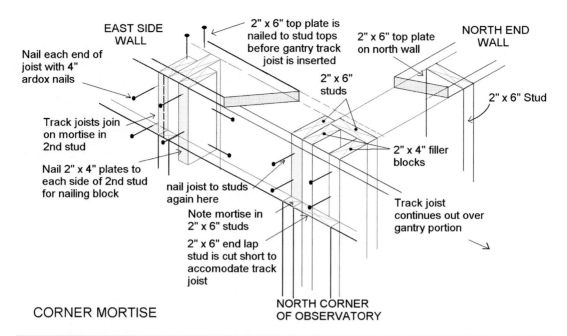

EAST SIDE WALL

2" x 6" top plate is nailed to stud tops before gantry track joist is inserted

2" x 6" top plate on north wall

NORTH END WALL

Nail each end of joist with 4" ardox nails

2" x 6" studs

2" x 6" Stud

Track joists join on mortise in 2nd stud

2" x 4" filler blocks

Nail 2" x 4" plates to each side of 2nd stud for nailing block

nail joist to studs again here

Track joist continues out over gantry portion

Note mortise in 2" x 6" studs

2" x 6" end lap stud is cut short to accomodate track joist

CORNER MORTISE

NORTH CORNER OF OBSERVATORY

Fig. 6.2 Detail of corner stud assembly showing mortise left in last two studs for the track joist (Diagram by John Hicks)

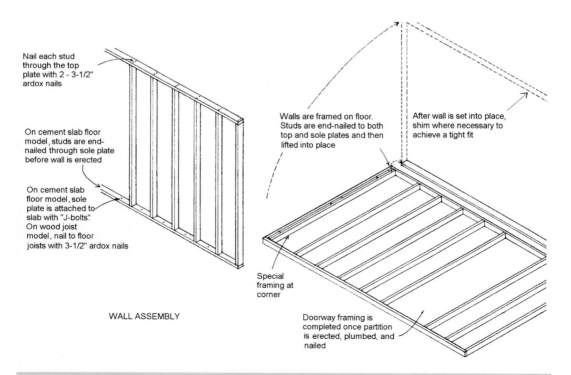

Nail each stud through the top plate with 2 - 3-1/2" ardox nails

On cement slab floor model, studs are end-nailed through sole plate before wall is erected

On cement slab floor model, sole plate is attached to slab with "J-bolts". On wood joist model, nail to floor joists with 3-1/2" ardox nails

Walls are framed on floor. Studs are end-nailed to both top and sole plates and then lifted into place

After wall is set into place, shim where necessary to achieve a tight fit

Special framing at corner

WALL ASSEMBLY

Doorway framing is completed once partition is erected, plumbed, and nailed

Fig. 6.3 Method of assembling and erecting the walls (Diagram adapted from Home Renovation—see (2) in Further Reading of Chap. 4)

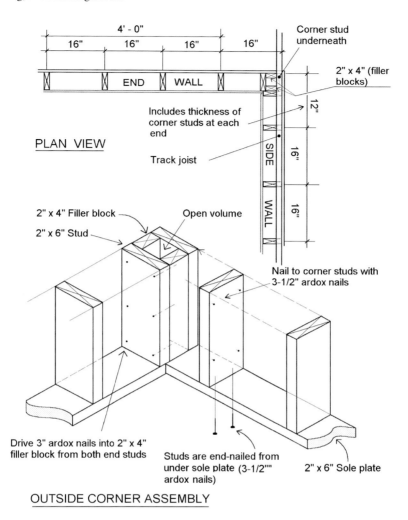

Fig. 6.4 Detail of outside corner assembly showing corner stud arrangement (Diagram adapted from Home Renovation—see (2) in Further Reading of Chap. 4)

Choosing, Locating, and Installing a Door

Choosing a Door

When framing the walls in our model leave a rough opening in the stud locations sufficient to locate your proposed door style. The dimensions for this opening, called a "rough opening" will be specified in your door literature. Once the walls are erected, you can cut out the sole plate in the rough opening. The door threshold will replace it. If you followed instructions earlier, there will be no j-bolt in this position. I advise you to purchase your door in advance so that you can allow for the exact rough opening you will require. I suggest the purchase of a steel panel door, with no window (for security

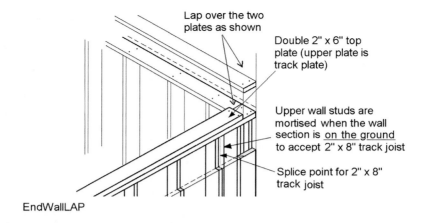

Lap over the two
plates as shown

Double 2" x 6" top
plate (upper plate is
track plate)

Upper wall studs are
mortised when the wall
section is on the ground
to accept 2" x 8" track joist

Splice point for 2" x 8"
track joist

EndWallLAP

Fig. 6.5 Detail of double top plate showing end-wall lap technique (Diagram adapted from Home Renovation — see (2) in Further Reading of Chap. 4)

and light-proofing), and a double lock set with no knobs but only dead bolts. Choose the stainless type if your budget allows, as the normal plated sets lose their finish quickly out in the elements. Make sure that the hinges have pressed-in pins (bearings) not the screw-in type which can easily be removed by vandals. A good security technique is to hammer a sturdy nail into the wood frame just above each of the hinges and drill holes in the door edge to seat the nails as the door is swung closed. No matter what a thief does to the hinges (even remove them completely) he cannot pull the door past the nails pinning it. Instead of a door knob that is easily broken into, install a door pull.

Locating the Door

The location for the door is a matter of individual preference, however if you plan to carry items from the observatory interior to the outside "patio" under the gantry, then it pays to place the door in the end wall. Normally this would be the North end wall. Place the door off to one side on the end, on the North-east end if the door swings inward so that when it is fully open it will lie flat against the east wall. If placed along a side wall, it is best to locate the door again in the North-east corner, again such that when it is swung inward it will lie this time against the North wall. Most of your observation will be in an arc from South-east through South to West, so don't put the door in this segment. Guests entering will either disrupt or block your line of sight and admit light in the wrong place entirely. A good tip is to remove the door after you have installed it, allowing for easier roof construction, adding interior sheathing, adjusting the pier etc. It's a nuisance while you are constructing the roof, particularly if you are climbing down to go outside and retrieve components. Make sure to allow for the door threshold.

The following diagram illustrates the pre-hung door assembly technique after the rough opening has been completed. The pre-hung door will have to be shimmed at various locations (or "blocked") to hold the pre-hung frame tightly. The diagram showing the blocking locations will help you determine where these "blocks" should be installed (Figs. 6.6 and 6.7).

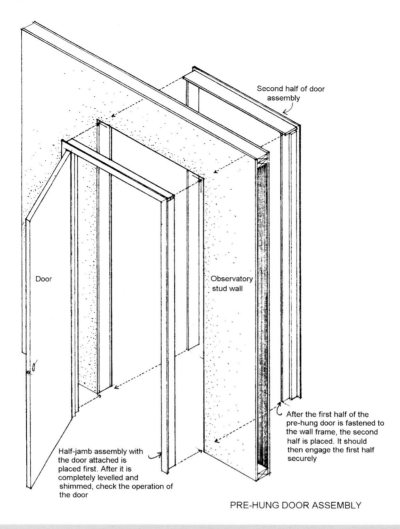

Second half of door assembly

Door

Observatory stud wall

After the first half of the pre-hung door is fastened to the wall frame, the second half is placed. It should then engage the first half securely

Half-jamb assembly with the door attached is placed first. After it is completely levelled and shimmed, check the operation of the door

PRE-HUNG DOOR ASSEMBLY

Fig. 6.6 Detail of pre-hung door assembly (Diagram adapted from Home Renovation — see (2) in Further Reading of Chap. 4)

Should you wish to construct the door and frame yourself the detail labeled "Self-hung door installation" will guide you in the construction of the jambs, stops, and blocking. The detail labeled "Head jamb, Side Jamb, and Sill" will assist you in understanding the cross-sectional make-up of the door. Finally, "Cylindrical lock and Strike plate installation," along with "Hinge and Knob installation" will complete the assembly. Is a tricky task for a beginner, and again I suggest you purchase a pre-hung door (Figs. 6.8, 6.9, 6.10, and 6.11).

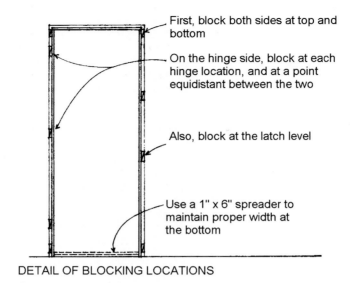

First, block both sides at top and bottom

On the hinge side, block at each hinge location, and at a point equidistant between the two

Also, block at the latch level

Use a 1" x 6" spreader to maintain proper width at the bottom

DETAIL OF BLOCKING LOCATIONS

Fig. 6.7 Detail of blocking locations (Diagram adapted from Home Renovation—see (2) in Further Reading of Chap. 4)

Installing the Metal Pier Top

Once the door (or at least the rough opening) is placed, the metal pier top section can be bolted down to its concrete base (if that was your chosen option instead of a full cement pier). It is better to wait until the walls are in place, and all top plates fastened before installing the pier top, as it would present an obstacle to positioning the walls, and could be damaged in the process. If you have not fabricated the metal pier top until now, take some measurements at this time to verify your earlier calculations for pier height. Measure the finished wall height from cement pier footing surface to the top surface of the wall plates, adding the estimated track gap which is 4½″ if using the 3″ diameter V-groove castors suggested. The total height should be 7 ft–4½″ if your total wall height (including top plates) is 7.0 ft. This is your clearance height maximum for a horizontal telescope. Subtracting the height of your telescope optical tube in horizontal position on its own mount from this height will give you an indication of your maximum metal pier height. Allow for some clearance, since the roof joists won't be all exactly at the same level, and there will be days you don't get the telescope tube completely horizontal on closing up (Fig. 6.12)!

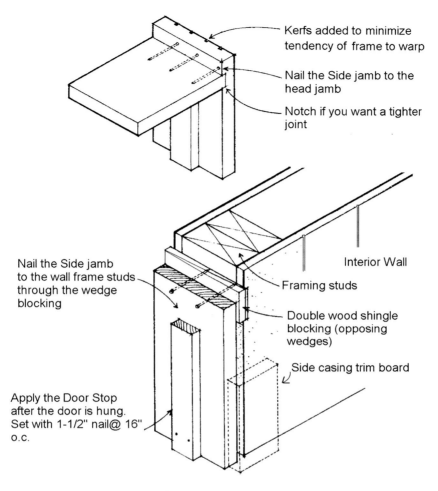

Kerfs added to minimize
tendency of frame to warp

Nail the Side jamb to the
head jamb

Notch if you want a tighter
joint

Nail the Side jamb
to the wall frame studs
through the wedge
blocking

Interior Wall

Framing studs

Double wood shingle
blocking (opposing
wedges)

Side casing trim board

Apply the Door Stop
after the door is hung.
Set with 1-1/2" nail@ 16"
o.c.

SELF - HUNG DOOR INSTALLATION

Fig. 6.8 Close-up detail of self-hung doorframe installation (Diagram adapted from Home Renovation—see (2) in Further Reading of Chap. 4)

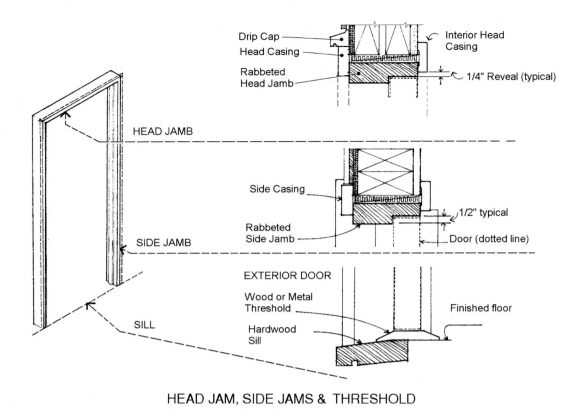

HEAD JAM, SIDE JAMS & THRESHOLD

Fig. 6.9 Detail of Head jamb, Side jamb and Threshold construction (Diagram adapted from Home Renovation—see (2) in Further Reading of Chap. 4)

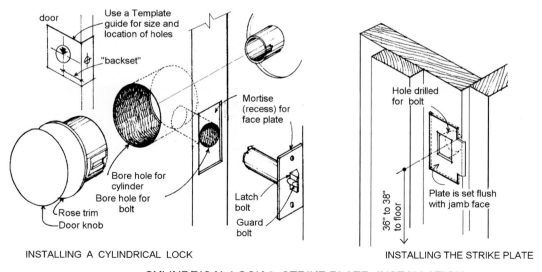

CYLINDRICAL LOCK & STRIKE PLATE INSTALLATION

Fig. 6.10 Detail of Cylindrical lock and Strike plate installation (Diagram adapted from Home Renovation—see (2) in Further Reading of Chap. 4)

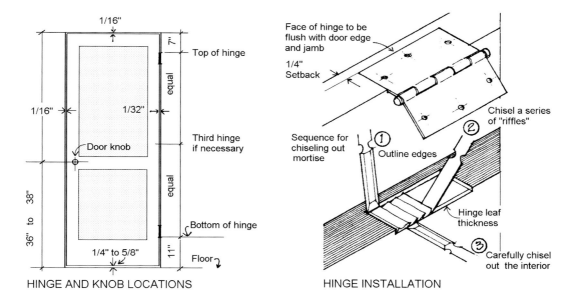

1/16"

Top of hinge

equal

1/16" → ←   1/32" → ←
← Door knob

7"

Third hinge
if necessary

equal

Bottom of hinge

36" to 38"

1/4" to 5/8"

11"

Floor

HINGE AND KNOB LOCATIONS

Face of hinge to be
flush with door edge
and jamb

1/4"
Setback

Sequence for
chiseling out
mortise

① Outline edges

②

Chisel a series
of "riffles"

Hinge leaf
thickness

③

Carefully chisel
out the interior

HINGE INSTALLATION

HINGE & KNOB INSTALLATION

Fig. 6.11 Detail of Hinge and Knob installation (Diagram adapted from Home Renovation—see (2) in Further Reading of Chap. 4)

Fig. 6.12 Photo—Guidescope on Mount intended for large refractor awaits final construction. The double upper plate can be clearly seen behind the mount (From the collection of John Hicks)

Chapter 7

Framing the Gantry Section

Whether you chose to install a full cement footing as an extension to the observatory floor, or you used sono tube piers, or a combination of the two, at this stage you will find it necessary to locate the gantry post supports precisely in their saddles.

To assure an effortless roll-off, the gantry and its track must be precisely in line with the observatory track section. This requires no deviation in height or angle, maintaining precise parallel supports for the steel castor track installed later. To fabricate such an arrangement requires careful attention to alignment and construction.

The first step involves temporarily placing the 6″ × 6″ posts in the four saddles now firmly set in concrete, and bracing them with 2″ × 4″ supports until each post is plumb (perfectly vertical). Use a carpenter's level to plumb the posts, then lightly nail the posts through the saddles just enough to hold them in temporary position.

When you are satisfied the framework of posts is square and "true," you can locate the position of the 2″ × 8″ track joist on the side of each post extending from the observatory proper. An easy way to achieve linearity in the track joist is with the use of a string line pinned to the top of the opposite end of the track joist in the observatory. Standing on a ladder beside the end post of the gantry, sight along the string pulled taught, and mark its intersection with the post. Providing the observatory track joist is perfectly level, you can also use a level on the new track joist to achieve linearity. Number the posts with a magic marker before taking them down, and locate them on your plan, so that each will go in its proper place in the final installation.

Take all the posts down and cut the mortise in them exactly as you did with the observatory studs. Be precise cutting the mortise so that the 2″ × 8″ track joist will fit flush with the surface of the 6″ × 6″ post and the 2″ × 6″ top plates will seat flat on top of post and joist (Fig. 7.1).

Re-install all the posts in their correct positions, and lag screw them through the saddle holes.

Use 5/16″ × 3″ long lag screws (don't use longer lag screws as they will meet in the center of the post).

Employ a washer with each lag screw and make sure both lag screws and washers are hot-dipped galvanized for endurance. Pre-drill a ¼″ guide hole for each lag screw (two lag screws are sufficient per post). Brace the posts as before with temporary 2″ × 4″ lumber. Install the 2″ × 8″ track joist in the

© Springer-Verlag New York 2016

J.S. Hicks, *Building a Roll-Off Roof or Dome Observatory*, The Patrick Moore
Practical Astronomy Series, DOI 10.1007/978-1-4939-3011-1_7

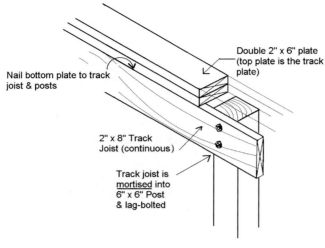

Nail bottom plate to track
joist & posts

Double 2" x 6" plate
(top plate is the track
plate)

2" x 8" Track
Joist (continuous)

Track joist is
mortised into
6" x 6" Post
& lag-bolted

POST, JOIST, AND TOP PLATES

Fig. 7.1 Detail—post mortised with track joist capped in double 2″×6″ plates. (Diagram by John Hicks)

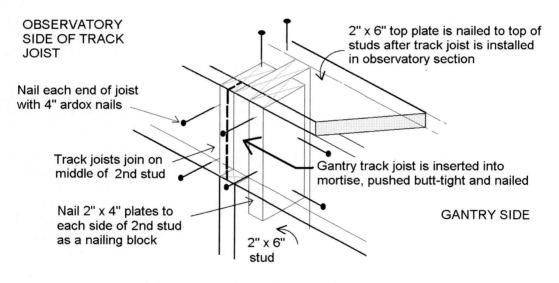

OBSERVATORY
SIDE OF TRACK
JOIST

2" x 6" top plate is nailed to top of
studs after track joist is installed
in observatory section

Nail each end of joist
with 4" ardox nails

Track joists join on
middle of 2nd stud

Gantry track joist is inserted into
mortise, pushed butt-tight and nailed

Nail 2" x 4" plates to
each side of 2nd stud
as a nailing block

GANTRY SIDE

2" x 6"
stud

JUNCTION POINT OF TRACK JOISTS

Fig. 7.2 Detail—junction of track joists inside the stud wall showing laminated nailing plates. (Diagram by John Hicks)

mortised posts, pushing the joist all the way into the observatory wall until it butts up against the joist in the stud wall (remember that we left the last stud gap of 16″ open for the gantry section of the track joist) (Fig. 7.2).

Lag screw the track joist to each post top with 5/16″×4″ hot-dipped galvanized lag screws—2 per post—set diagonally). Pre-drill all lag screw guide holes with a ¼″ drill. In the observatory wall section, use two 4″ ardox nails per stud, nailed flush so that outside panels will remain flat when nailed over them. Note the nailing plates laminated to each side of the stud where the joists meet mid-way on the stud. Nail the joist into each of these plates using two 4″ ardox nails. Make sure to lag screw the nailing plates on to the stud leaving room for the ardox nails to penetrate.

Leave a full 16 ft of track joist extending from the observatory wall to beyond the last post—this will be exactly what is needed to hold the entire roof rolled completely off the observatory.

Finally, nail on the double 2″×6″ top plate in line with the double top plate inside the observatory.

Nail on the first 2″×6″ plate first with 4″ ardox nails about 16″ apart into the track joist maintaining it flush with the joist's outer surface, and into the tops of the posts. Then nail the upper 2″×6″ plate exactly over this, making sure that no joints are co-incident (maintains rigidity). Use 3½″ nails about every 16″ (same as the observatory section). This effectively forms a 3½″×5½″ beam over the 2″×8″ track joist creating a strong support for the roof once rolled off the walls of the observatory (refer to Fig. 7.1).

Chapter 8

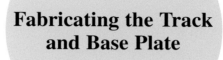

Fabricating the Track and Base Plate

There are a multitude of roll-off mechanisms in use. Some are simple to construct but they usually require more mechanical advantage to operate. Rubber wheeled casters, for example, flatten from prolonged periods of inactivity and hobble along the track base requiring a lot of force to move them. This is made worse by colder temperatures. A winch system consisting of stainless cable, various pulleys and a lot of preliminary "engineering" solves the problem. Success isn't usually achieved on the first attempt either because of the requirement for very specific locations of the pulleys. On one observatory I designed, the owner had intended to use a "closed-loop" winch system but was forced to resort to a two-winch system, one to winch off the roof, and another to winch it back on. The system worked well although he installed the two winches on pedestals mounted on the floor, which took up needed space within the observatory. Some very southern owners may even find that in order to see Polaris and polar-align their telescopes they may have to extend the gantry further north enabling them to pull the roof a greater distance on the track than it would normally travel. In observatory construction, some builders suggest using golf balls entrapped in a wooden or metal channel, but they will rattle along awkwardly as they "gang" up and roll in a convoy. They inevitably clash, grate together, and produce an unacceptable amount of friction. I have seen examples of golf balls contained by various tubing, or channel to lessen the frictional problem, but in the final analysis it is difficult to prevent them from rubbing together and creating a drag.

Our design uses five V-groove steel casters on each side of the rolling roof. These are typically used to roll on the back of an inverted angle iron rail, rather than on a "flat" leg section of the angle, although both will work. As we shall see later on, if the angle is laid flat with the caster running on a flat side its "leg-up" side will produce unwanted friction. The model I specify is made by Bestway Casters in Toronto, Ontario, with the designation "V-groove Steel Caster 3½″ × 1½″ Model No. 1303-VG-RB." The caster has roller bearings and rigid forks. They are rugged casters which will last the lifetime of your observatory. If you are unable to purchase from Bestway, present your dealer with the above specification, which will guarantee an approximate match in strength and endurance (Figs. 8.1, 8.2, and 8.3).

The track, all of 30 ft long on each side, is fabricated with 1½″ inverted steel angle, nip-welded to a continuous 3½″ wide × 3/16″ min thick steel plate. You can make the plate 5½″ wide to cover the entire surface of the wood track plate underneath if you prefer but no less in thickness. It is best to

© Springer-Verlag New York 2016

J.S. Hicks, *Building a Roll-Off Roof or Dome Observatory*, The Patrick Moore
Practical Astronomy Series, DOI 10.1007/978-1-4939-3011-1_8

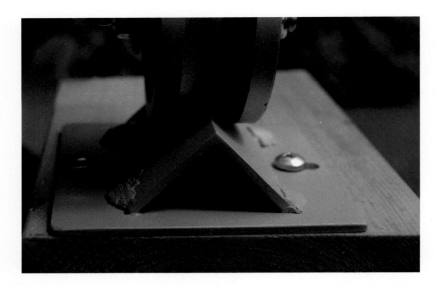

Fig. 8.1 Photo—V-Groove caster on section of inverted angle resting on double 2″×6″ plate. (Photo by John Hicks)

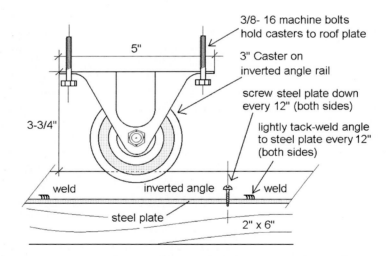

Fig. 8.2 Detail—side view V-Groove caster resting on inverted angle track. (Diagram by John Hicks)

make the steel base plate in as long a section as transportation will permit. If you can persuade a metal fabricator to assemble it in 15 ft sections it would be best, although hard to carry without bending, and you should increase the thickness of the plate to ¼″. Such a length requires a long flat bed truck to transport it and two men to carry it. For self-transport, six 10 ft sections would be next in preference. If your car or van only holds 8 ft lengths of lumber, then six 8 ft sections plus two 6 ft sections will suffice. Up to four sections on each track introduces three joints for the casters to roll over, but if you reverse the sequence of lengths in laying the track, one side relative to the other, there will be no common joint for the casters to run over. This may be a bonus in the long run and a good reason to use three or four lengths of track over two equal ones on each side.

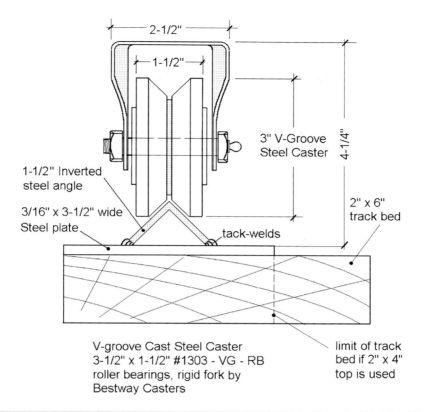

Fig. 8.3 Detail—end view V-Groove caster resting on inverted angle track. (Diagram by John Hicks)

Ask the metal fabricator to make slotted holes on both sides of the steel base plate about every foot, such that adjustments can be made in widening or narrowing the width of the track simply by loosening the hold-down screws and sliding the track to the left or right on the wooden plate underneath (the uppermost 2″×6″). The slotted holes should accommodate ¼″ wide fasteners and be well-finished so that a 1/4″ diameter screw will slide easily in the slot (Fig. 8.4a).

The inverted angle is nip-welded to the steel base plate, in its center, about every 12″ on both sides of the angle. Intense welding is to be discouraged as it will warp the track and steel plate. Make sure to ask the fabricator to cut off the angle lengths cleanly and squarely, such that when the rails are butted together it's a flush joint—with no gaps in the top rail junction. When the track is finished, spray it twice with a good coat of metal primer and allow it to dry thoroughly. Finally, spray with two coats of Tremclad or similar rust proof coating in the color of your choice. I like to exhibit the mechanical workings to visitors, so I coat the tracks in bright orange. If you want to conceal what they are, spray the tracks in a black or dark blue.

Installing the Track

Position the tracks on the observatory section first, starting from the southernmost observatory wall. Use zinc-coated 1½″ long × 1/4″diameter large Robertson wood screws, but do not screw the tracks down tightly at this time. Initially, the base plates should locate such that their outer edges are flush

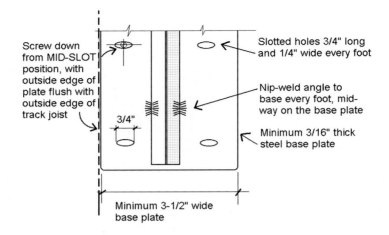

Screw down from MID-SLOT position, with outside edge of plate flush with outside edge of track joist

3/4"

Slotted holes 3/4" long and 1/4" wide every foot

Nip-weld angle to base every foot, midway on the base plate

Minimum 3/16" thick steel base plate

Minimum 3-1/2" wide base plate

FABRICATING THE STEEL TRACK BASE PLATE

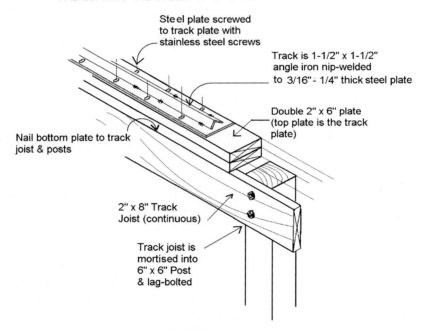

Steel plate screwed to track plate with stainless steel screws

Track is 1-1/2" x 1-1/2" angle iron nip-welded to 3/16" - 1/4" thick steel plate

Double 2" x 6" plate (top plate is the track plate)

Nail bottom plate to track joist & posts

2" x 8" Track Joist (continuous)

Track joist is mortised into 6" x 6" Post & lag-bolted

POST, TRACK BEAM AND TRACK CONNECTION

Fig. 8.4 (**a**) Detail—plan view of track section showing nip-welds and slotted holes. (Diagram by John Hicks) (**b**) Detail of post with track joist, track plate, with steel track in place. (Diagram by John Hicks)

with the outer face of the $2'' \times 8''$ track joists, and in this position the screws should be mid-way in their slots. This will allow a slight relocation of the track base should it be necessary. The effects of shrinking wood or a foundation movement could produce just enough shift in the track to change the critical distance between tracks. Butt each of the track sections up firmly against each other such that the casters won't encounter any misaligned or rough joints. Make sure that the ends at each joint are square and clean (you may have to file them smooth). Measuring progressively across from the top of the angle on one side to the top of the angle on the other, install the track across the observatory walls out over the gantry making sure the separation is constant (Fig. 8.4b).

Alternative Track Designs

The Flat Track Rail

The inverted angle track could be replaced by a $1'' \times 2'' \times 3/16''$ thick angle positioned on the $2'' \times 6''$ track bed with its $1''$ leg up (Figs. 8.5 and 8.6).

Providing that the screws holding down the angle to the $2'' \times 6''$ bed are recessed sufficiently, the casters will roll over the flat leg of the rail with moderate ease, although more "bumpy" than the V-groove-inverted rail set-up. There are two flaws in this arrangement. The $1''$ leg-up side will rub on the outside face of the caster wheels producing unwanted friction as the wheels contact it, and the recessed flat-head screws will create a rough road-bed. The design offers additional security however, in that the $1''$ leg-up prevents any jump-off the caster rail, confining the casters within the two track angles. The observatory I designed in Florida utilized this arrangement, but required a winch system to roll the roof off and roll it back into position.

The Garage Door Track and Roller Alternative

Although designed for a lighter purpose, the garage/barn door type roller and track will support a heavier roof load if it contains enough casters. The problems with the system are twofold: the track is light and because it "contains" the caster tends to increase the frictional area, and the joints in the track are more difficult to align. Using this type of system will require a sturdy beam on top of the stud walls to support the track because it will now be held vertically on the inside face of the beam (the track is vertically-oriented). The casters ride in prefabricated carriages bolted to the underside of the roof plate. Since the casters have to "reach" down into the track on the inside of the studs, an extra

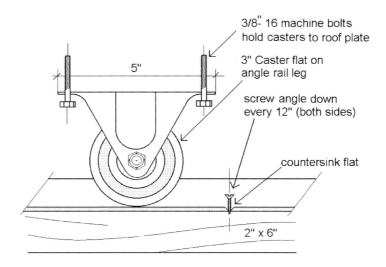

Fig. 8.5 Detail—side view V-Groove caster running on flat leg of angle with leg-up. (Diagram by John Hicks)

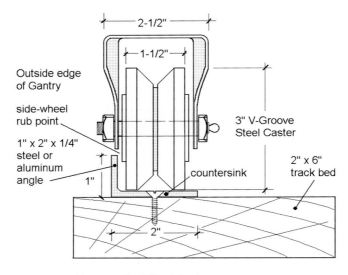

Fig. 8.6 Detail—end view V-Groove caster running on flat leg of angle with leg-up. (Diagram by John Hicks)

Fig. 8.7 Photo—Terry Ussher "Sky Shed" Observatory showing garage door track and rollers on double plate under roof (Courtesy of Terry Ussher)

$4'' \times 4''$ beam must also be bolted underside the roof plate (or a double $2'' \times 6''$ plate). This type of caster arrangement also has an added advantage: it reduces the air-gap between roof and wall edges because the casters "reach down" into a vertical track fixed to the inside wall edge. This seats the roof joists lower on the wall, dropping the soffit considerably (Fig. 8.7).

Although there are a number of alternative roller systems, the V-groove caster and inverted angle offers the best strength and ease of movement.

Track Ends and Stops

Before advancing to the roof fabrication, it is wise to install the Caster stops on the ends of both tracks. The caster wheel stops are simply made from a 12″ long block of 4″×4″ pressure-treated timber that has one end cut at 90° and the other at 45° (for design purposes only). It is beneficial to install a 1/4″ slab of rubber on the 90° ends of the stops for cushioning, as the heavy roof could do some damage if pushed back too forcefully. The caster stops are pre-drilled for long lag screws which bed down the stops on the track plate-top plate assembly. Put them at the very end of the track—each end, and on both sides. The casters should be placed to allow a 12″ overhang on each end of the track at the track stop (i.e. when the roof is at rest over the observatory in fully closed position, the roof soffits should extend 12″ further over each end (Fig. 8.8).

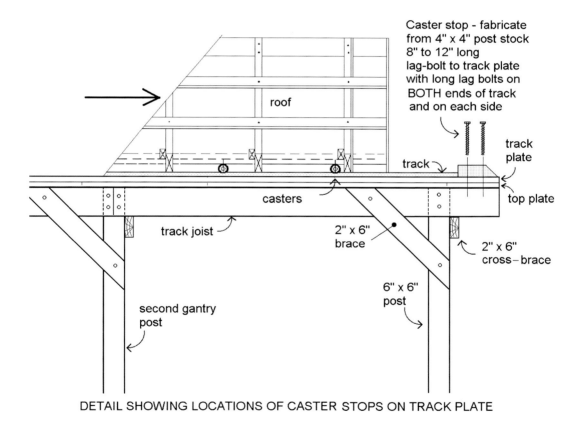

DETAIL SHOWING LOCATIONS OF CASTER STOPS ON TRACK PLATE

Fig. 8.8 Detail showing locations of caster stops on track plate. (Diagram by John Hicks)

Chapter 9

Framing the Roof

The Assembly Process

The roof should be framed in place, meaning that the caster plate (the bottom plate of the roof framework) is placed on the track with casters intact, and the roof trusses built upon it. Otherwise, the roof could be constructed on a "bench" of sawhorses and lifted into place. This is an extremely difficult operation without the assistance of a crane, because the walls of the observatory (including foundation) are too high for a crew of people to simply lift the roof over and onto the track. Unless you have an army of helpers enlisted to do the job, a group of people cannot lift so much weight over their heads. If you have crane access (the mobile type) that won't alter the appearance of your lawn, the saw horse "bench" process is the way to go, saving you the arduous task of climbing up a step ladder each and every time you add a framed truss, to nail it in place. Later, when installing the 15 pound roof felt underlay and the corrugated steel roof panels, the requirement for reaching up into the roof framework becomes necessary. This becomes particularly difficult when applying the roof felt and later the steel roof sheets starting at the ridge board, although you can get nailing access through the openings in the trusses and purlins (the metal roof sheets are normally about 30 in. wide).

Construction of the Caster Plate

Generally then, it is advisable to construct the caster plate first, bolting on all the casters in exact position. This has the added advantage of being able to "test" the casters on the track, making sure they all roll without binding and lie perfectly in line on the caster plate. Once fabricated, the caster plates can be held together across the span of the gantry or the observatory with $2'' \times 4''$ braces. Usually the roof is constructed over the observatory section because the floor is higher than the ground level outside in the gantry section, and it is smoother and easier to work upon. You will drop a myriad of fasteners and parts by accident, easily retrieved on the observatory floor, rather than

© Springer-Verlag New York 2016
J.S. Hicks, *Building a Roll-Off Roof or Dome Observatory*, The Patrick Moore
Practical Astronomy Series, DOI 10.1007/978-1-4939-3011-1_9

N.B. Bolt casters down to caster rail first before installing
roof trusses on it (much easier than working upside-down)

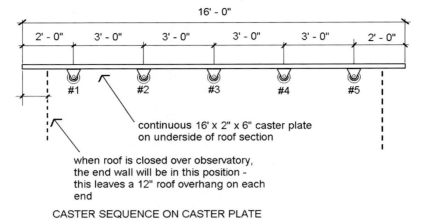

CASTER SEQUENCE ON CASTER PLATE

Fig. 9.1 Detail of caster sequence and alignment on the 2″ × 6″ caster plate (Diagram by John Hicks)

searching for them in gravel or bare soil under the gantry. The following diagram illustrates how to position the casters on the caster plate (Fig. 9.1).

Bolt the casters in place right through the caster plate precisely, paying particular attention to the center-line of the V-groove as you place each caster in its position. The brand of caster I recommend does not have slotted holes in the caster base which means you cannot re-adjust their position laterally. You have to get it right the first time. I recommend using a string line fixed just over the groove in the end caster, pulling it tight over each caster installed to check for any misalignment. If you had an extra 16 ft section of track available, that would make a perfect "gauge" for testing their alignment as you proceed.

This is unlikely, however due to the added cost.

Calculating the Rafter Lengths

Typically, a minimum requirement in roof pitch is 3 in 12 (3 in. vertical rise per 12″ of run) for medium snow loads. Our design uses 6 in. rise in 12 in. of run, which is more than sufficient for northern cold climates with heavier snow loads. This works out to a rise of 3 ft over a 6 ft span which is ½ the roof width. To calculate what length of rafters you will need, think of the vertical rise as the vertical side of a right triangle, and the run as the horizontal side, then calculate the hypotenuse (or rafter length). In our case this becomes 6″ 8½″ (6 ft 8½″ in.), derived from 3 ft squared plus 6 ft squared = 45, where the square root of 45 is 6.708 ft (or 6 ft 8½″).

From this length deduct 1/2 the ridge board thickness (1/2 of 1–1½″ = 3/4″) to arrive at the main rafter length from ridge to the outside face of the observatory wall [i.e. 6 ft 8½″ − 3/4″ = 6 ft 7¾″ (6 ft 7¾″ in.)].

Since there is a 12 in. overhang for the soffit, the same right triangle calculation will produce a hypotenuse of 13½″ i.e. 6 in. squared plus 12 in. squared = 180 in. The square root of 180 is 13.416 in. or 13½″ (close enough). Adding this to the rafter length above, we arrive at (6 ft 7¾″ + 13½″ = 7 ft 9¼″

total rafter length. One should add an inch at each end to allow for an angled "plumb" and "tail" cut in each rafter. Total length is then 7 ft 11¼″ (7 ft 11¼ in.). This is convenient because you will be cutting your rafters from 8 ft stock and have very little waste lumber.

Framing Sequence (Layout of the Rafters on the Caster Plate)

Since there is no "bird's mouth" in the bottom end of the rafter it is not necessary to locate one, making your rafter installation a lot less precise than it could have been. A long "bottom chord" plus "king posts" hold the rafters in place quite rigidly. Mark the "plumb cut" at the top (ridge board contact) and the "tail cut" at the bottom of the rafter with an inverted framing square (flip the square over) and cut the rafter at these points. The top mark is located by positioning the framing square *at the top end of the rafter* (see detail 9.2), aligning the 6 in. mark on the tongue (short leg of the square) and the 12 in. mark on the body *with the top edge of the rafter*. Scribe or mark a line along the tongue of the square, this will be the center of the ridge pole. Then move the square down in the same 6/12 orientation exactly ¾″ (½ the thickness of the ridge board). Mark the "Plumb Cut" point here. Measure down the rafter 6 ft 7¾″ to the point where the rafter intersects with the caster plate (the outside edge of the wall below) and scribe a mark there.

From that point measure 13½″ (the distance to the end of the overhang) and using the inverted framing square, mark off the "tail cut" (see detail 9.2). Cut the rafter at these marks (plumb cut and tail cut).

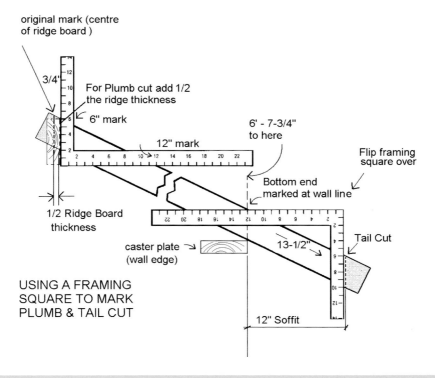

Fig. 9.2 Detail showing method of marking "tail cut" and "plumb cut" on rafter (Diagram adapted from Creative Homeowner Press) (1)

You can also measure down the entire rafter from the "plumb cut line" with the framing square alone by "stepping off " six increments of 12″ by holding the square in the 6/12 orientation on the rafter. The six increments will locate the intersection with the caster plate (the outside edge of the wall below). Mark this point, and then invert the square using the 6/12 orientation, and measure the last 12″ of horizontal overhang. Cut the rafter here. (See details in Figs. 9.2 and 9.3).

Cut the mortises in the rafter for the roof purlins at this stage by sawing the 1½″ depth and chiseling out the 3½″ width of a 2″×4″ if you go this route (recommended). Alternatively, framing anchors can be used without mortises with short between-rafter purlins but the assembly is not as rigid as a mortise joint and the full-length purlins. Refer to the following details (Figs. 9.4 and 9.5).

STEP 1

Lay out and assemble one roof framing member section first, including the two rafters, a section of ridge board and the bottom chord, cut to the specifications outlined above. Nail it all firmly together. This will be used as a template to make all the eight trusses. Use a short piece of 2″×6″ plank to duplicate the ridge board thickness and temporarily nail it in place (do not set the nails, but leave heads protruding—you will have to remove them to fit the real ridge board when in place permanently).

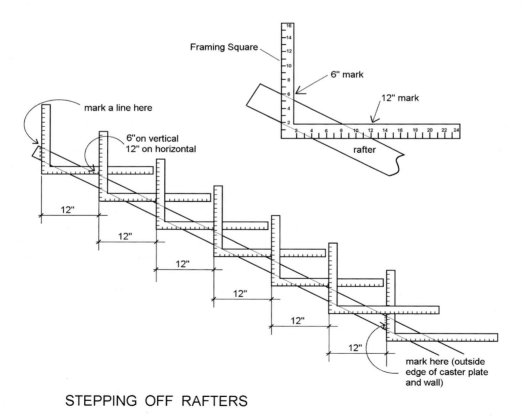

STEPPING OFF RAFTERS

Fig. 9.3 Detail showing method of "stepping off" 12″ horizontal increments from top to bottom of rafter (Diagram adapted from Creative Homeowner Press—see ref. 1)

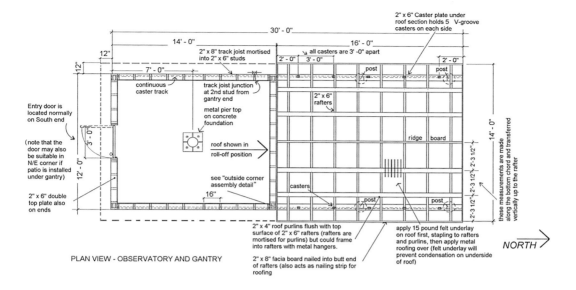

Fig. 9.4 Plan view detail of entire observatory showing roof section and purlin spacing measured on the slope of the rafter (Diagram by John Hicks)

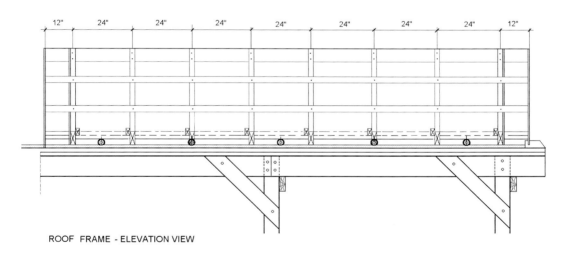

Fig. 9.5 Elevation of entire roof over gantry showing the rafter spacing with purlins (Diagram by John Hicks)

Attach the $2'' \times 4''$ king post supporting the ridge board from the bottom chord. There is a convenient metal tie to fasten these two items together called a Simpson Hurricane/Seismic tie, or you can use a regular joist hanger, nailing the bottom flange of it to the king post flank. Either way it's a lot easier than mortising the king post into the ridge board (Fig. 9.6).

You will also have to temporarily install a short section of $2'' \times 4''$ at the bottom for the continuous chord running the length of the roof. Set the template on your caster plate and tie it in temporarily with $2'' \times 4''$ struts. Make any adjustments to the length of the rafters at this time, if any (Fig. 9.7).

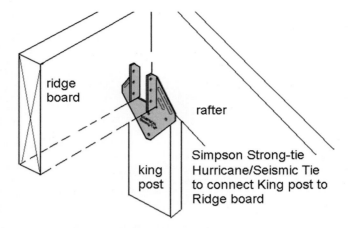

Fig. 9.6 Detail of "Simpson Hurricane/Seismic-tie" used to connect King Post to Ridge Board (Diagram by John Hicks)

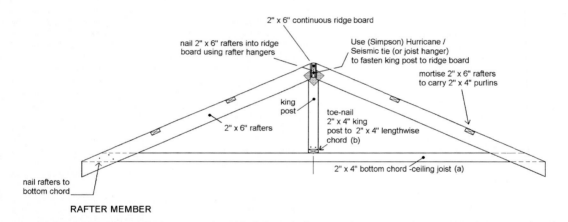

Fig. 9.7 Roof framing member detail showing rafters, purlins, ridge board, bottom chord, king post, and continuous chord (Diagram by John Hicks)

STEP 2

Cut the ridge board to length and mark the rafter locations on it every 24″ O.C. so you know exactly where to place them. Mark the positions for the rafters also on the caster plate. Make sure to allow for the 1 ft overhang at the gable end when you cut the ridge board.

STEP 3

Once you have verified the exact length of the rafters and the bottom chord after placing the "test" rafter frame on the caster plate, take it down and construct seven more pairs exactly the same length. On the ground make another complete framing member. This time, attach the two rafter members to

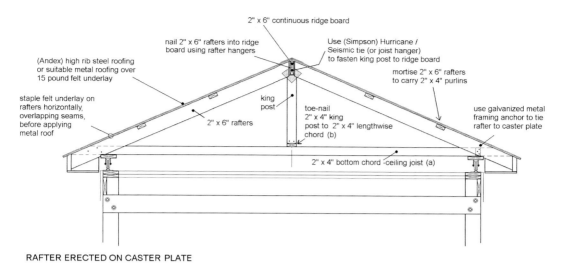

RAFTER ERECTED ON CASTER PLATE

Fig. 9.8 Detail showing roof framing member in place on caster plate (Diagram by John Hicks)

the ridge board in the position of "gable ends". Use either the "Hurricane-ties" or joist-hangers to fix the rafters to the ridge board. Following this, tie the two rafters on each frame with their bottom chords (a) and insert their king posts resting on the continuous $2'' \times 4''$ chord (b) running length-wise. At this point you will have constructed the two end gables held together by the ridge board, and the $2'' \times 4''$ chord under the king post (see the following diagram) (Fig. 9.8).

STEP 4

Get some help to lift this assembly up onto the track, and attach temporary braces to the end walls to support the assembly. Braces can be sections of $2'' \times 6''$ running up the studs of both end walls, temporarily nailed into both end gable members. Tie each rafter section down to the caster plate with a framing anchor on the open side of the rafter with the bottom chord nailed flush on the other side (Fig. 9.9).

Fill in all the remaining rafter frames and frame in the gable ends using a ladder to get access to the roof now in place. Install the $2'' \times 4''$ purlins, making sure the frames are plumb (Fig. 9.10).

Following the basic erection of the roof framework, the next task should be to fabricate the boxed-in-soffits before the roof sheathing is applied, which is a lot easier done without the roof deck in the way. Cut triangular sections from $2'' \times 6''$ which will fit under your rafter overhang. You will find that a triangular section $6'' \times 12''$ will suffice giving you a hypotenuse of $13\frac{1}{2}''$ which will run under the rafter. These are called "look-outs". Cut 16 "lookouts" in total for the eight pairs of rafter-members. Glue these under the rafter end, flush to the $2'' \times 8''$ fascia board which will nail to the end of the rafter. Also, employ metal hangers to attach them to the joists since glue will only last so long. After the glue has dried cut the plywood soffit panels and allow room for a standard "vent-strip" available at lumber supply centers. Screw these in place after you have nailed on the $2'' \times 8''$ fascia board across the ends of the rafters. This board will likely need planing to fit the profile of the roof edge unless you wish to leave a small lip below the soffit panels (Fig. 9.11). Leave ½" air space between the soffit plywood and the observatory walls.

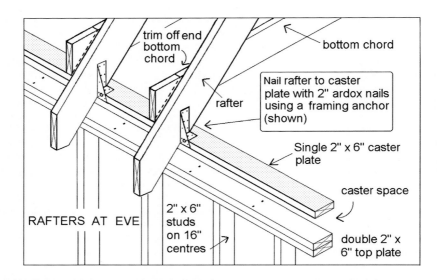

trim off end bottom chord

bottom chord

Nail rafter to caster plate with 2" ardox nails using a framing anchor (shown)

rafter

Single 2" x 6" caster plate

caster space

RAFTERS AT EVE

2" x 6" studs on 16" centres

double 2" x 6" top plate

Fig. 9.9 Detail showing rafter-caster plate connection and use of framing anchors (Diagram by John Hicks)

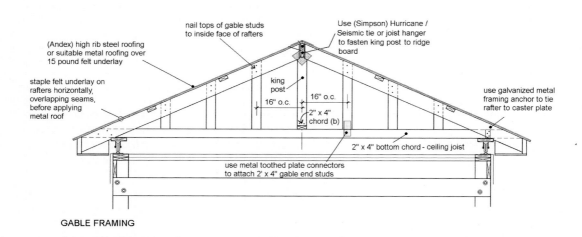

nail tops of gable studs to inside face of rafters

Use (Simpson) Hurricane / Seismic tie or joist hanger to fasten king post to ridge board

(Andex) high rib steel roofing or suitable metal roofing over 15 pound felt underlay

staple felt underlay on rafters horizontally, overlapping seams, before applying metal roof

king post

16" o.c.

16" o.c.

2" x 4" chord (b)

use galvanized metal framing anchor to tie rafter to caster plate

2" x 4" bottom chord - ceiling joist

use metal toothed plate connectors to attach 2' x 4" gable end studs

GABLE FRAMING

Fig. 9.10 Detail of gable end showing installation of $2'' \times 4''$ gable studs (Diagram by John Hicks)

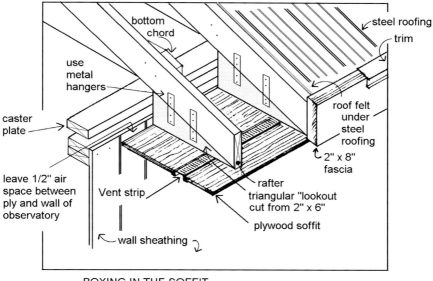

BOXING IN THE SOFFIT

Fig. 9.11 Detail showing method of boxing in the soffit (Diagram adapted from Home Renovation—see (2) in Further Reading of Chap. 4)

Further Reading

1. Barrett James (1993) The Concise Guide To Building Sheds, Creative Homeowner Press, A Division of Federal Marketing Corp., Upper Saddle River, New Jersey, USA

Chapter 10

Applying the Steel Roofing

The choice in roofing falls logically to sheet steel roofing manufactured in panels and cut to length for you by the manufacturer. Requiring no roof sheathing for our application other than roof felt stapled onto the purlins and rafters (since it does not need to be insulated), the application of metal roofing directly over the roof frame saves weight, while outlasting shingles and almost all other normal roofing. The diagrams earlier in this book portray a roof frame with only two rows of purlins on each side of the ridge board: that would only be sufficient should you use standard roof sheathing (ply deck) and asphalt shingles.

If you choose to apply shingles you need to apply 3/8″ or 7/16″ plywood sheathing first over the entire roof framework, then apply your choice of shingles. This type of roofing will be much heavier than the steel panels, and I would not advise it.

For the application of steel roofing you will also need an extra run of 2″×4″ purlins—one row about 2 ft from the ridge board, and another row about 2 ft up from the fascia board. These short 24″ sections of 2″×4″ purlins can be installed with prefabricated metal hangers made for 2″×4″ studs. Nailed flush with the top edge of the rafter, the hangers hold the purlins quite rigidly. The top edges and bottom edges of the steel sheets will screw down to these short purlins, supplying an attachment point at top and bottom of the rafters. The reason I did not specify the prefab metal hangers earlier was for purposes of structural rigidity—the majority of long, mortised, purlins tie the roof together very well, whereas the hanger-type connections can provide some "wiggle" on a large, heavy movable roof (Fig. 10.1).

Measure your roof surface across from gable to gable and ridge board to eves, calculating the area required. Steel roofing suppliers will normally supply you with all the fasteners required for the area you dictate to them. These are hardened screws with neoprene washers attached, designed to be driven with a power driver. They will also supply the ridge cap, rake flashing (for along the outside edge of the gables), and eve flashing. You will require all of these components. You will not need the closure strips (or seals) since the roof isn't insulated and you want air-flow through it anyway. Your first action is to measure the distance from the top of the ridge to the eve, and order the exact length of steel roof sheet that fits. The company will supply it cut cleanly. If you have to cut it with a steel

© Springer-Verlag New York 2016

J.S. Hicks, *Building a Roll-Off Roof or Dome Observatory*, The Patrick Moore
Practical Astronomy Series, DOI 10.1007/978-1-4939-3011-1_10

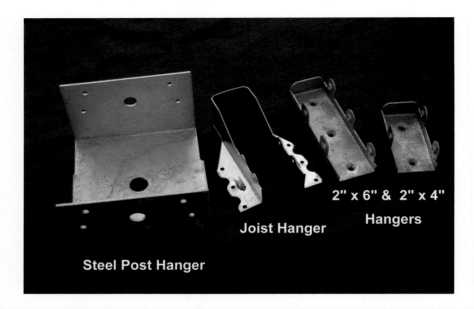

Fig. 10.1 Photo of various sheet metal hangers for joists, purlins, post saddle (Photo by John Hicks)

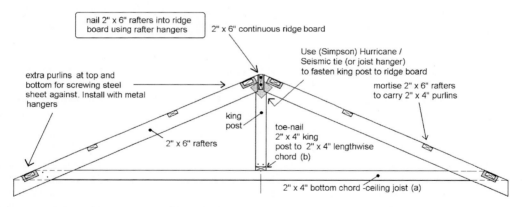

EXTRA PURLINS, FRAMED INTO RAFTERS WITH METAL HANGERS

Fig. 10.2 Section of rafter showing required extra purlins for sheet steel attachment at ridge and eaves (Diagram by John Hicks)

cutting saw, it will be messy. I suggest a panel design called "*Century Rib*" which has a repetitive mix of larger and smaller standing "ribs." The larger "rib" on the outside edge of the sheet overlaps the next larger "rib" on the adjacent panel being applied, helping to hold it while you screw it down (Figs. 10.2 and 10.3).

Fig. 10.3 Photo of a typical steel roof sheet—"Century Rib" design by Andex Metal Products (Photo by John Hicks)

Tools and Equipment Required

The following selection of tools would normally be required to install steel panel roofing.

- Hammer
- Electric drill for pre-drilling holes in sheets for fasteners (better than punching)
- Cordless Screw driver and bit to match roof screws
- Steel cutting disk on a Skill saw
- Shears
- Tape measure
- Leather gloves
- Chalk line
- Safety glasses (always wear them, and especially when cutting steel)
- vice grips
- rope

Always use gloves while handling or working with steel roof sheets as the edges are sharp and produce nasty, deep cuts. When cutting with the Skill saw and metal disk, watch for sparks, and always wear the safety glasses

Applying the Sheet Steel

STEP 1 Installing Felt Underlay

The first step involves the nailing down of 15 lb felt underlay horizontally across the roof members. Use galvanized shingle nails and pull the felt underlay tight while applying it rafter-to-rafter. Make sure to provide about a 2″ overlap on each row of felt starting from the roof ridge downward over the roof. Do not apply the felt over the ridge board since it will seal off air flow from the interior of the observatory, rising to the ridge cap. Its purpose will be to conduct "hot" observatory air out of roof volume accumulating all day in the sun.

STEP 2 Install Eave Flashing

Screw on the two Eave flashings first, they will cover a small edge of roof surface and the fascia board (they are pre-bent for this purpose). You must put this on first as you cannot get it under the steel sheets later, once they are screwed down. Do not put on the Gable trim now because it overlaps the top edge of the last steel sheet you install at the edge of the gable.

STEP 3 Test for Squaring

Check the roof squareness. At the corner where an eve and gable meet, measure 8 ft back along the roof edge at the eve and mark it. Then measure 6 ft up the gable roof edge from the eave and make another mark. Measure the distance between the two marks. If you find that it is exactly 10 ft, your roof is square at that particular corner. Checking all the roof corners will verify "squareness."

STEP 4 Installing Roof sheets

If the roof is not square you can correct this on the overlap joints of the steel roofing, little by little, by taking advantage of the tolerances in the laps. Also at the end gable, the gable flashing (rake) will hide any imperfections as far as "slant". To place the steel sheets up on the roof, put a board against the fascia board and slope it gradually as possible to the ground surface. Clamp a vice grip to the upper edge of the sheet, attach a rope and pull it up onto the roof. Be careful of wind, and prevent the sheet from buckling. Start at the eave furthest from the prevailing wind. Extend the sheet about 1″ over the gable edge and 1″–2″ over the eave. Do not fasten the leading edge furthest from the gable edge before starting the second row because the second row sheet must overlap the first. The overlap point for screwing the two sheets down must always occur on the large standing rib nearest the edge of the sheet. Leave a space at the ridge to allow a little ventilation under the ridge cap. *The sheets should stop short of the ridge.* Drill through both sheets at the overlapping junction of sheets such that one roofing screw will penetrate and tie down both. When you arrive at the other gable side, test-fit the last sheet, mark the gable end, and take the sheet down to cut it (allow an extra 1″ over the gable end Facia board as before). Use a steel-cutting disc on a standard Skill saw, with gloves and **safety glasses on**—sparks will fly!

STEP 5 Install Ridge Cap

Place the ridge cap on the ridge and make sure its position is even on both sides of the roof. Mark the edges with a chalk line or felt pen. When fastening the ridge cap, drive the screws through the large standing ribs on the roof sheets (not the lower ribs on the sheet because the ridge cap metal will pucker between the raised ribs making an unsightly appearance). For a ridge longer than the length of ridge cap supplied, overlap the two caps slightly.

STEP 6 Install Gable Trim

The gable trim is the last to be installed. The best style overlaps the steel roof at the edge of the gable extending down over the fascia board. Make sure you specify the same color for all the flashings, trim, and ridge cap. A color different from the roofing color looks horrible.

Check before you install it because outdoor light can be deceiving on the shiny steel surfaces. Your roof when completed will make you feel very proud of your work as it will appear structurally perfect.

Further Reading

1. Canadian Sheet Steel Building Institute, CSSBI, How to Series – Light Gauge Steel Roofing & Siding, Cambridge, Ontario, Canada

Chapter 11

Applying Exterior Siding and Interior Wall sheathing

There is a great variety of exterior siding materials available. The normal exterior wall covering (at least the underlay), is usually plywood or waferboard. This can be installed directly over the studs, vertically. If you plan to cover this in a wood-board type siding, several designs are available to you. Most have a 1 inch nominal thickness (actually 3/4 inch thick) and are usually installed vertically. If you are planning to use wood board siding you will have to install blocking at midspan between the wall studs to nail the boards to, since you would only have the sole plate at the bottom and the double top plates at the top.

Typical patterns which suit observatories are "channel groove siding" and "board-and-batten" style siding. The following detail illustrates the two types and the basic application technique for each (Fig. 11.1).

Sometimes local building codes will require sheathing under the board siding i.e. like plywood or waferboard. Hardboard siding (manufactured from compressed wood pulp) is available in a variety of traditional board designs in $4' \times 8'$ panels. These are fairly long-lasting and can mimic the real board texture quite well, complete with a simulated wood grain. My design incorporates vinyl siding because I did not want another out-building to stain or paint periodically. I also prefer the vinyl siding because insects such as wasps, find it impossible to build nests on. Animal control is also a must in the observatory with sensitive computer cables to chew on etc., and the board siding is easy to climb up on and "squeeze" into the soffit space. Vinyl is too slippery for most squirrels especially if you run a coating of "Armour-all" over the siding periodically.

© Springer-Verlag New York 2016
J.S. Hicks, *Building a Roll-Off Roof or Dome Observatory*, The Patrick Moore
Practical Astronomy Series, DOI 10.1007/978-1-4939-3011-1_11

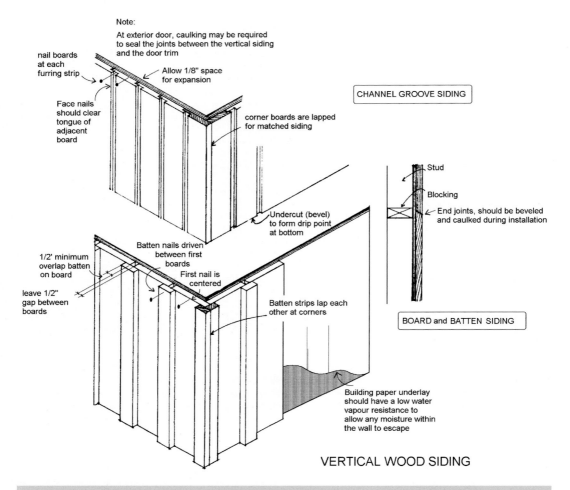

Note:
At exterior door, caulking may be required
to seal the joints between the vertical siding
and the door trim

nail boards
at each
furring strip

Allow 1/8" space
for expansion

CHANNEL GROOVE SIDING

Face nails
should clear
tongue of
adjacent
board

corner boards are lapped
for matched siding

Stud

Blocking

End joints, should be beveled
and caulked during installation

Undercut (bevel)
to form drip point
at bottom

1/2' minimum
overlap batten
on board

Batten nails driven
between first
boards

First nail is
centered

Batten strips lap each
other at corners

BOARD and BATTEN SIDING

leave 1/2"
gap between
boards

Building paper underlay
should have a low water
vapour resistance to
allow any moisture within
the wall to escape

VERTICAL WOOD SIDING

Fig. 11.1 Detail illustrating the application technique for Channel Groove siding and Board and Batten Siding (Diagram adapted from Home Renovation—see (2) in Further Reading of Chap. 4)

Vinyl siding comes in a variety of "earthy" colors which match environmental surroundings quite well, fitting into your landscape or rear yard without dominating it. The vinyl "package" comes with trim for doors and corners, and is quite easy to apply, nailed down with color-matching nails. It does require an underlay of ply or waferboard, but this doesn't have to be applied seamlessly since the vinyl will cover all defects in application of the underlay.

You must consult your local aluminum siding supplier for this material. The great thing about vinyl is that you never need to paint it (Fig. 11.2).

Gable Vents and Fan Systems

It is very advantageous to have an exhaust fan on at least one gable end wall to exhaust the hot air accumulating all day in the closed observatory. So much the better if it is either timed or on a thermostat. This helps your cool-down time for any observing session, night or day. A minimum requirement would be the installation of two air vent grills on both gable wall ends without a fan, as this

Fig. 11.2 Photo of vinyl siding application on domed observatory exhibiting its elegant appearance (Photo by John Hicks)

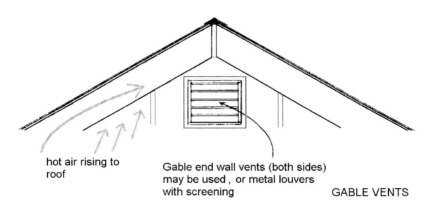

hot air rising to roof

Gable end wall vents (both sides) may be used , or metal louvers with screening

GABLE VENTS

Fig. 11.3 Detail of Gable Vent air flow (Diagram adapted from Home Renovation— see (2) in Further Reading of Chap. 4)

arrangement allows a passive air flow-through. Most wall fans are designed to fit within studs on a stud wall, and the kits come with full installation instructions. The simplest procedure is to cut exact apertures in the exterior wall board, if wood, and place the fan housing beside the king post in the gable (the highest point in the wall of your observatory), to remove the most heated air.

Vinyl siding is a little bit harder to cut once applied, so it is best to install the fan first in this case and cut the vinyl to fit around it snugly. A vent application will require some pre-engineered cutting also. You will also have to pre-wire the fans (if you use fans) through the wall stud openings to an electrical box on the inside wall of the observatory, before you install the interior wall sheathing (Figs. 11.3, 11.4, and 11.5).

Fig. 11.4 Photo of gable vent on Ussher Observatory (Courtesy of Terry Ussher)

Fig. 11.5 Photo of internal fan housing fitted to inside of gable wall (Courtesy of Jay Ballauer)

Interior Wall Sheathing

Most observatories I have inspected are painted dull black inside to eliminate any reflected light. Thus, the wall texture really doesn't have to be anything special since it almost vanishes in the night. Standard 3/8 inch ply sheathing would be quite acceptable, with narrow 1-1/2 inch x 1/4 inch combing strips nailed over the seams. Since I am a solar observer specialist, I prefer the white glossy walls

Fig. 11.6 Looking down into the interior of my domed observatory at the white masonite panels and the aluminum strips over the seams. Notice how high the pier is in this high-wall observatory (Photo by John Hicks)

that you see in the earlier photos of Don Trombino's Observatory (The Davis Memorial Observatory). This effect was achieved by applying pre-coated gloss-white masonite panels directly over the studs. I finished the seams in bands of anodized 1½ inch × 1/8 inch aluminum strips. These were fastened with stainless steel round-head screws and stainless finishing washers. The effect was truly professional. An extra bonus is that I can remove them for access to wiring etc., without using a pry-bar and mutilating the wall surfaces. Anything you place in the observatory such as a lap-top computer, small table etc., fits well within this interior wall scheme. It is however, not suitable for the night sky purist whose observations will be mostly in the dark. If you prefer the texture of masonite panels which are hard and quite thin, you can pre-coat the clear finished masonite with a starch wash which will help to bind flat black paint on its surface. For example you could use the same technique that I used with aluminum strips on the seams sprayed flat black. The effect would be the same. Figure 11.6 shows the neat effect created by masonite panel sheathing and aluminum seam strips.

Chapter 12

Final Luxuries

A Winch System for Opening and Closing the Roof

The following photo illustrates a well-designed winch system for removal and closure of the roll-off roof, placed on the observatory floor against the north wall (Fig. 12.1).

The steel cable for such a system has to be strong. The type of cable used for sailboat stays is ideal, usually 1/8 inch - 3/16 inch diameter. You will need at least 100 ft of stainless steel cable for this mechanism. The "rolling-off" winch cable will attach to the inside of the south <u>roof gable</u>, then north to a pulley on the end <u>cross-member</u> of the gantry. From there it runs back south across the gantry and over a pillow- block (or "fairlead") fastened to the top plate of the observatory section north wall. Bending over the pillow block, the cable runs down the north wall interior and terminates on a winch installed on the wall, or on a floor-pedestal. The "closing-roof" cable will run from another winch, up the north wall interior, over another pillow-block, to fasten on the inside of the north <u>roof gable</u>, preferably at its bottom chord for strength. The system requires two winches in order that the wires do not get snagged, and also two pillow blocks so that the wires do not run over each other on the blocks and bind. Cranking the "closing-roof" winch simply winds the cable up, pulling the roof back over the observatory. Fasten the end of the cable to the face of the bottom north chord of the rolling roof section so that it will be pulled directly over the pillow-block in line with the inside north wall of the observatory. This allows the soffit to overhang the wall as usual. Winding onecable will, of course, cause the other to unwind so a light brake of some kind is necessary, or the wire will uncoil off the winch in a rather unruly manner. It works well, and can be motorized as the following photo illustrates. Battery operated, the 12 V car/boat winch is bolted to the floor with a control box next to it for forward or backward motion of the pulley, which translates into roll- off, or roll-on motion. The system is tricky to engineer, and much "fiddling around" with the tension, pulley locations, is necessary before final satisfaction. Notice that the motorized winch system is bolted right into the cement floor in Don Trombino's observatory . He altered the initial "winch-operated" system to a 2-pulley on one shaft electric-motor-driven system, engineering a "tension" system on both pulleys to prevent the cable from unwinding on the floor (Fig. 12.2).

© Springer-Verlag New York 2016
J.S. Hicks, *Building a Roll-Off Roof or Dome Observatory*, The Patrick Moore
Practical Astronomy Series, DOI 10.1007/978-1-4939-3011-1_12

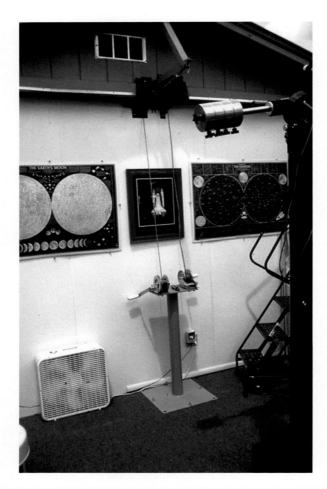

Fig. 12.1 A double winch system allows both roll-off and roll-back-on functions to be done within the observatory. (From the collection of John Hicks)

Comfort Items

Most observatory floors, particularly cement slab type, should be covered in a carpet to prevent lenses and other equipment from breaking when dropped, as they surely will be. It is customary to use household living room carpet, and is usually achieved when the homeowner re-carpets the home and has a quantity of old carpet available. It pays to go to a carpet supplier/installer and ask if he can supply you with a section of used, but acceptable, carpet for your observatory. In my observatory it simply lies on the floor, fitted exactly to the confines of the walls. Being heavy enough not to curl or slide it was never glued or fastened to the cement floor which is rough enough to grip it. Installing carpet over a plywood floor on joists will require stapling or nailing. The Pier is also covered for warmth with carpet cut to exactly wrap around it. It takes a while to sew it together along a vertical seam but it is the best way to attach it tightly. This comes in handy when you are viewing at the zenith

Fig. 12.2 Motorized winch system operating on a 12 V battery. (Photo from the collection of John Hicks)

close-up to the pier, if you are using a refractor on a German equatorial mount which demands that you come close to the pier. Cold cement or steel is not a welcome sensation against your legs when you are observing or tracking celestial objects. The use of carpet also "softens" the "mechanical aspect" of the observatory, and it offers yet another bonus—when you are awaiting an event, or just relaxed from a long observing session you can lie on the floor and take in the glory of the heavens above you (Fig. 12.3).

I also find the use of a "rolling ladder" for access to the instrument, and to the roof, to be an advantage. Often I need to re-position a pulley, or to remove an insect nest, and the rolling ladder is handy. It also serves as a great "perched" desk for my lap-top, which easily fits with mouse, mouse pad, and log book on the uppermost platform-stair of the ladder right beside the telescope. I often sit on this platform at the level of the telescope (mine was perched on a very high 7 ft pier) and just watch the sky overhead.

It is on summer days, after viewing the active surface of the Sun, which is the mainstay of my hobby, that I feel so much joy in having an observatory. It is almost a religious experience to lie on the warm carpeted floor and watch the beauty of the summer sky from your own "camera obscura." It is really a "cloister" of sorts, where you can undertake your study in isolation, where you can concentrate free of the hub-hub in the world around you.

In building this observatory, you will discover a magical sphere of activity from the minute you mark out its "footprint" on the ground to the moment you install carpet on its finished floor. The "magic" begins when you survey the contents of this book, visualizing what your observatory will look like out in your yard, and continues through every hour you spend building this structure. It will not be easy, but rewarding every time you leave it after a day of construction, satisfied with your progress, and dreaming of the day it is finished. I guarantee you will be swept away, on a journey that will build self confidence and satisfaction, developing skills you may never have felt you possessed. For sure, your neighbors' curiosity will be perked, and his/her respect for your inventiveness will escalate. The Universe lies waiting for you to experience it, don't waste another day.

Fig. 12.3 Rug-covered pier offers real "creature warmth" in the "machinery" of the observatory. (Photo by John Hicks)

Part II
The Dome Observatory

Chapter 13

Dome Design Materials and Construction Methods

Design and Construction Considerations

Many astronomers have a preference for a dome observatory, but feel that building one is beyond their skills because of the curved surfaces required. Some extra effort may be required to produce a dome, but there are no special skills required, nor equipment, beyond the use of most home workshop power tools. The metal track, arches, ribs, and dome base ring should be contracted out to a local metal fabricator, who will have the machinery to bend the materials into perfect circular sections. The only extra equipment required to build the hemispheric dome involves the use of a band-saw for cutting out the laminated plywood rings that serve as castor ring under the dome, plus the circular sole plate and top plate, forming the top and bottom of the base walls. If buying a band-saw seems to be too much of a burden, these components could also be contracted out to a local wood-working shop—as I did. I preferred to work on forming the pier, the cement floor, and fabricating the metal components of the pier while someone else was doing all the brute work, cutting out numerous plywood segments and bending the thick aluminum framework of the dome. Besides, this hurried construction along with two other craftsmen working on separate tasks together. The base of the observatory is entirely wood—frame construction with the exception of the steel track and pre-hung steel entry door. The prototype for the dome observatory in this book is my own model with an aluminum dome and a 7 ft (2100 mm) high-wall base—tall enough to accommodate a full-size door. In earlier attempts at construction, I built a 4 ft (1200 mm) low-wall base, which required me to stoop, almost crawl, through the low door frame. After several years of striking my head and damaging equipment carried through the low opening, I designed, and built, the present 7 ft (2100 mm) high-wall model (Figs. 13.1 and 13.2).

Walking through a full-size steel door is also much more acceptable to senior astronomers who lack the flexibility to crawl through low places. (Think ahead, because an observatory will last you well into your senior years, particularly if the entry door is accessible and the dome is made of metal

© Springer-Verlag New York 2016
J.S. Hicks, *Building a Roll-Off Roof or Dome Observatory*, The Patrick Moore
Practical Astronomy Series, DOI 10.1007/978-1-4939-3011-1_13

Fig. 13.1 My first dome observatory, a 4 ft (1200 mm) low-wall model, housed an 8 in. (200 mm) Schmidt-Cassegrain (a short-barrel mirror telescope)

Fig. 13.2 My second dome observatory, a 7 ft (2100 mm) model, was built for long focus refractors, and is the design I recommend in this book

(aluminum, galvalume or galvanized steel). If you wish to construct your dome observatory lower, the construction techniques are the same with a much lower pier, and less over-all cost. The 4 ft low-wall model will place your upper torso within the dome itself, eliminating the need for a rolling ladder to reach the telescope. It has other benefits: the average height astronomer will be able to reach most of the upper dome interior with a small step-ladder, allowing for the construction of the dome

Fig. 13.3 Jack Newton guiding his finished dome in place while it is lifted by a crane at his Arizona Sky Village home property [photo courtesy of Jack Newton]

on top of the base, along with future painting and any repairs of the track, casters, and the slot door opening/closing mechanism. The high-wall model dome observatory must be entirely built on the ground, with the completed dome lifted up onto the track when work is entirely finished. This is best accomplished with a light crane, or with a group of willing astronomers who must lift it over their heads and place it on the track—a hazardous operation which can only be done before the skirt is added , since it prevents lifting the dome onto the track underneath it (Fig. 13.3).

Once perched 7 ft high on its track, the dome is almost impossible to access with a ladder, if you dare to use one, for it will slide easily off its curved surface. Consider also the type of instrument you are housing, or the one that you ultimately wish to own, because long focus refractors require both a low and high swing clearance, particularly when pointing towards the zenith. If you contemplate future planetary work, and think you may want to purchase a long focal length refractor, its eyepiece will be swung almost down to floor level on a low pier, low-wall model, requiring you to observe bent-over near the cold floor. Obviously the high-wall model serves your best needs in this case. A wide aperture, low-focal-length Newtonian reflector or a Schmidt-Cassegrain type instrument suits a low-wall model, because of its shorter tube, or its eyepiece located at a convenient eye-level. Because I undertake a lot of solar astronomy involving long focus refractors with added optical components, the higher pier, higher dome, was an absolute necessity. Spend some time before construction considering which wall height will suit your present and future needs.

Commonly Used Materials and Construction Techniques

The most common design owned by amateur astronomers in the past was the free-standing circular base with a rotating dome on top. There are now many variations to both the dome and the base of the observatory including the materials used to fabricate them. Some builders have utilized the common farm silo dome seen scattered throughout the rural landscape, cutting open the dome center to create an observing slot, and re-enforcing its edges with structural arches underneath.

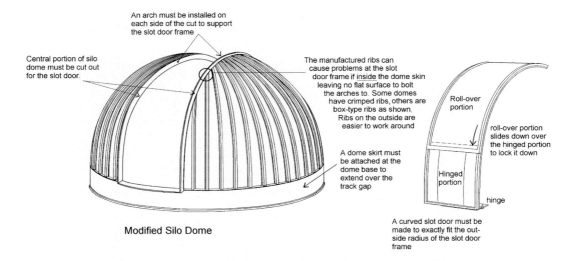

An arch must be installed on
each side of the cut to support
the slot door frame

Central portion of silo
dome must be cut out
for the slot door.

The manufactured ribs can
cause problems at the slot
door frame if <u>inside</u> the dome skin
leaving no flat surface to bolt
the arches to. Some domes
have crimped ribs, others are
box-type ribs as shown.
Ribs on the outside are
easier to work around

Roll-over
portion

roll-over portion
slides down over
the hinged portion
to lock it down

A dome skirt must
be attached at the
dome base to
extend over the
track gap

Hinged
portion

hinge

Modified Silo Dome

A curved slot door must be
made to exactly fit the outside radius of the slot door
frame

Fig. 13.4 Diagrammatic view of the required cut into the silo dome and slot door that must be built to fit over it. Note that the top portion slides up an over the dome releasing the bottom hinged portion (when closed, the bottom few inches of the upper slot door cover the top few inches of the hinged lower portion serving as a lock-down) [diagram by John Hicks]

Although this may at first appear to be an easy way to create a dome, it is not as easy as one might imagine. Farm silos are usually made of galvanized steel, and often 'galvalune' which is an aluminum-galvanized steel blend. The pre-rolled, curved panels (or gores) are held together in a dome by flanged seams, sometimes formed into each panel as a "standing rib" or "box rib", or held together with clips, depending upon the manufacturer. These eliminate the need for the fabrication and assembly of ribs to hold the gore panels together. The pre-manufactured domes most always have a circular, flat, crown section capping the dome such that the gore panels do not converge to a point at the top. The caps usually cover too large an area at the top of the dome, wider in circumference than the slot door width required. Once the slot door area is cut out of the dome, the residual openings left by the cap must be filled in (Figs. 13.4 and 13.5).

Once the dome is cut open to create the observing slot, arches must be installed on either side of the opening to support the slot door frame and the slot door itself. This is best done before the slot is cut out with a steel saw since it prevents flexure of the panels while cutting is undertaken. The arches should be bolted to the steel panels throughout their curvature, their edges guiding the saw as it cuts through the panels and their seams.

If these seams (standing rib, box rib or crimped) are raised on the outside of the dome, it is easy to bolt the arches inside the dome, and to fit the slot door frame later onto the arches (Figs. 13.6, 13.7, 13.8, 13.9, and 13.10).

For most dome builders, the common choice of materials are wood, masonite, or fiberglass—all easier to work with than steel or aluminum. A variety of rib framework materials have been used to support the dome panels, ranging from plywood laminates, tubes, and aluminum bars, curved to form a hemispherical dome. Since laminated wood plywood is easy to work with, it is most often the material chosen for the ribs, arches, dome and base rings. Fastening the panels to the ribs is relatively easy with hand tools and stainless screws. The problem is always the effect of weather on wood laminates that deteriorate rapidly, leaking or rotting unless constant attention is given to maintenance (caulking, sealing, and painting). Some builders use masonite or thin door veneer for dome panels, and others utilize home-made molds to cast curved fiberglass panels which readily form a perfect dome. Thin plywood panels can be cut with "kerfs" on the inside allowing them to bend, but if the "kerfs" are

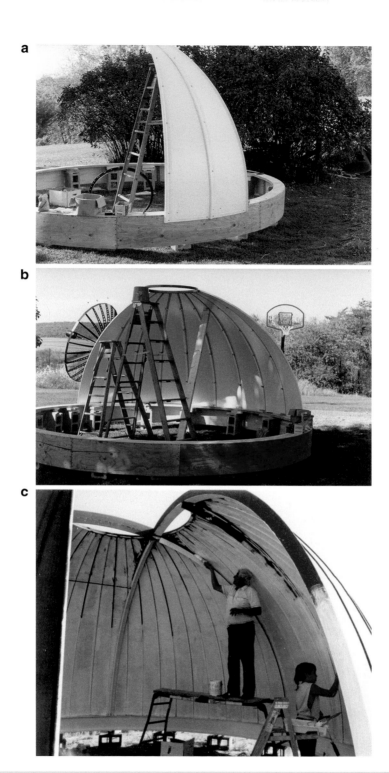

Fig. 13.5 (**a**) Calvin Cassel's silo dome observatory showing the application of the first two gore panels onto the plywood castor ring [photo by Calvin Cassel Jr.]. (**b**) Calvin Cassel's silo dome observatory showing the gore panel application well under way. Note the assembly is propped up by a 2″ × 4″ strut to keep the unsupported panels standing in place. The original ring tying the panels together at the top is also visible [photo by Calvin Cassel Jr.]. (**c**) View of a silo dome observatory under construction showing the circular cut-out left by removing the dome cap and the installation of the slot door arches. Note that the arch (one of two) extends right over the full hemisphere of the dome [photo by Alan Otterson, Albuquerque Astronomical Society]

Fig. 13.6 Outside view of dome before installation of the slot door and painting. The observatory base is shown under construction nearby [photo by Alan Otterson, Albuquerque Astronomical Society]

Fig. 13.7 View of the inside top of the finished silo dome observatory showing the circular cap area filled in, and the mechanical crank system for moving the slot door over the dome [photo by Alan Otterson, Abuquerque Astronomical Society]

Fig. 13.8 View of inside the finished silo dome showing the two arches extending down the backside of the dome. Note the large, long focal length Newtonian Reflector [photo by Alan Otterson, Albuquerque Astronomical Society]

Fig. 13.9 View of interior Dome retaining roller used to keep the dome from falling off its track [photo by Alan Otterson, Albuquerque Astronomical Society]

Fig. 13.10 (a) View of finished General Nathan Twining Observatory [photo by Alan Otterson, Albuquerque Astronomical Society]. (b) View of Calvin Cassel Jr. finished observatory on its round base [photo by Calvin Cassel Jr.]

too deep, the outer side of the plywood panels may crack and weather quickly. Two part epoxy coatings, normally used for boat hulls, can be applied over the panels and exposed parts of the dome arches to provide a long-lasting, weather-resistant finish. Most of the wood products used (plywood, masonite, door veneer) resist bending in two directions due to their rigid nature or thickness. Light gauge aluminum panels of the same gauge supplied in standard house exterior cladding can be forced to bend in two directions with some effort, producing a long-lasting, professional-looking dome (Figs. 13.11, 13.12, 13.13, 13.14, and 13.15).

Fig. 13.11 Arches, ribs, base rings and caster ring can be cut from plywood sheet, glued and screwed together to form curved segments—structural support for the dome

Fig. 13.12 Thin plywood door veneer can be successfully bent to form the dome gore panels. Plywood up to 1/8″ (3 mm) thick can be cut with narrow saw cuts or "kerfs" on the inside to allow it to bend easier. The weakened "kerfed" surface can be painted with epoxy cement to re-establish its rigidity

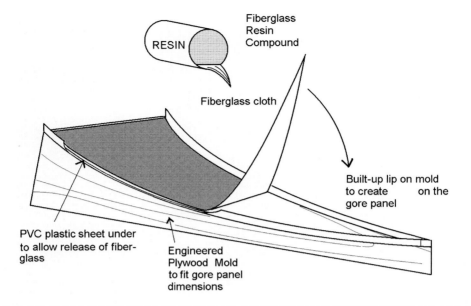

Fig. 13.13 Making a mold for gore panels in fiberglass resin and cloth is an excellent idea. After several coats of resin have been applied over fiberglass cloth, the exterior of the panel can be coated with a two-part epoxy white finish producing a smooth dome exterior

Fig. 13.14 A finished veneer dome showing the slot door overlap

Fig. 13.15 Jack Newton's domes stand out clearly against the mountain background at his Arizona Sky Village home. These domes are all made from a plywood framework of arches, and ribs covered in door veneer as in Figs. 13.11 and 13.12. The left hand dome was built by AVS resident Rick Beno and the right hand dome by Jack Newton [photo by Jack Newton]

An Outstanding Dome Design

An odd variation to the normal dome design is a dome whose panels do not converge toward the apex of the dome. Such a dome was designed and built by my old friend Walter Wrightman. Walter's vision of a dome involved one with ribs that went right over the dome perpendicular to the base much like the slices in a loaf of round bread. His observatory, built with plywood arches, ribs, and dome ring, covered in a skin of thick aluminum panel, was truly a novel design I had never seen attempted before (Fig. 13.16).

After visiting me and examining the fabrication of my aluminum dome, Walter struggled with a way to eliminate the compound bending required in the normal gore panels made from aluminum skin. Walter solved the problem of forcing the aluminum skin to bend in two directions by designing a panel that did not taper to a point, but bent in one direction only, completely over the dome hemisphere. This eliminated the troublesome buckling of the aluminum sheet at the rib junctions where rivets held the panels in place.

In building my dome, the application of the gore panels was a learning process that was made easier by starting the rivets in the central area of the panel, alternating the riveting toward the top and bottom of the panel as I progressed. It was a tedious process because the panel had to be initially pre-fitted, then secured flush with the dome base ring at the bottom so that it fit perfectly in place at the top, resting on the ribs all the way up. Walter's 'hemispheric slices' solved that problem but introduced another. In order for Walter's panels to fit comfortably on the plywood ribs and arches,

Fig. 13.16 View of Walter's unique observatory on a low-wall base behind his home in Newmarket Ontario [photo by Walter Wrightman]

Fig. 13.17 View of Walter's dome showing structural components, all assembled from laminated plywood members, fastened together with sections of aluminum angle [photo by Walter Wrightman]

the laminated ribs and arches underneath had to be planed to conform to the curvature of the dome. This required the use of an electric plane to form the edges of ribs and arches exactly to the intended radius of the final dome. Testing the beveling with a plywood template of inside radius equal to the intended dome radius, Walter kept all the beveling uniform. To build a dome in this fashion, an electric plane is a necessity, and will be an extra piece of equipment required (Fig. 13.17).

1. Dome with standard ribs converging
to a point at their apex

Shape of standard gore panels
cut to fit beween ribs - note the
bulge in all sides to facilitate
the compound bend required

Fig. 13.18 Diagram illustrating gore panels converging to a centroid on a conventional Dome. Alongside is a typical shape of a gore panel—curved slightly on all sides to accommodate the bending required [diagram by John Hicks]

2. Dome with parallel ribs in
non-converging "slices"

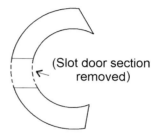

(Slot door section
removed)

A typical lower panel in the "sliced" rib
model takes on a bizzare shape with
the slot door section removed

Fig. 13.19 Diagram illustrating Walter's vertical 'sliced bread' panels that extend right over the hemisphere of the dome. Alongside is a typical panel, which has a bizarre shape in order to fit 'over-the-dome' [diagram by John Hicks]

Walter was a superb craftsman, and working with wood was easier for him than bending aluminum structural components. His aluminum panels were cut to match 1/8″ (3 mm) thick cardboard templates that he had pre-fitted to the rib sections, eliminating any error in measurement. Using a thicker gauge aluminum than I applied on my dome (because his panels were only required to bend in one direction), he drilled holes through the aluminum panels into the plywood ribs and arches, securing them together with stainless screws. In addition to this, he overlapped each panel with its neighbour (from each side of the dome toward the top), assuring a shingle-like weather-proofing. In this way, only a single row of screws fixed adjacent panels to a single rib underneath. It stands today as an excellent design, providing you have the patience to plane the entire dome framework into a smooth hemisphere. It also offers the opportunity to use thicker aluminum skin for the panels. Walter had solved a difficult construction problem to produce a very durable dome (Figs. 13.18, 13.19, and 13.20).

The Warm Room Addition

If the height of the observatory dome above ground is sufficient enough, a small rectangular addition can be added, joined to the dome walls at mid-diameter (Figs. 13.21 and 13.22).

This is a design option that dome observatory builders should consider for a warm computer station, or work room space. The warm room blends into the dome observatory flush with its curved wall and extends outward 10 ft (3.0 m). Note that if a larger warm room is required, its length can be

Fig. 13.20 View of Walter's original dome fitted on a new elevated base and pier by Jeffrey Boylin at his 'Sky Dome' observatory. The original observatory base was elevated higher by placing it on sono-tube piers with a new floor. Elevating the old base achieved two purposes: it eliminated pouring a thick cement foundation and allowed easier access to the inside by stepping up and into the observatory [photo by Jeff Boylin]

increased, with the effect of altering the roof slope, and lowering its drainage potential. The warm room is a luxury for any astronomer, housing a desk, chair, and bookcase for reference texts or equipment storage. Almost every astronomer, sooner or later, will attempt some astrophotography, requiring a computer station, with warmth and comfort to enjoy the frigid nights. In Fig. 13.22 it would seem that the minimum height for the observatory wall (under the dome) must be at least 10′ (3 m) in order to provide enough roof slope, ceiling height, and space for a staircase in the warm room. The limiting parameter appears to be the height and position of the observing floor which must satisfy both space for the stairwell below it, and the operational requirements of the telescope above it. To assure maximum safety on the observing floor, a short trap door is necessary to avoid consuming too much floor space and to provide an easy access from below. The 5′–6″ (1670 mm) long trap door was chosen to allow sufficient head room up through the stairwell but not to intrude excessively into the observatory floor. It could be raised by a small winch and pulley system located in the warm room, to rest against the pier when opened. In larger diameter domes, the stairwell could be left permanently open, protected by a railing to leave sufficient clearance around the telescope for all observing positions. However, the internal door connecting the warm room to the observatory would have to be closed during observing sessions, as it seals off the space created by the stairwell access, blocking warm air rising from the warm room. This design places the observing floor 4′–6″ (1370 mm) above the concrete slab floor, and approximately 5′–6″ (1675 mm) below the laminated plywood top plate of the observatory wall. Adding the track gap plus the distance to the sill of the slot door, the interval becomes more like 6′–6″ (1980 mm).

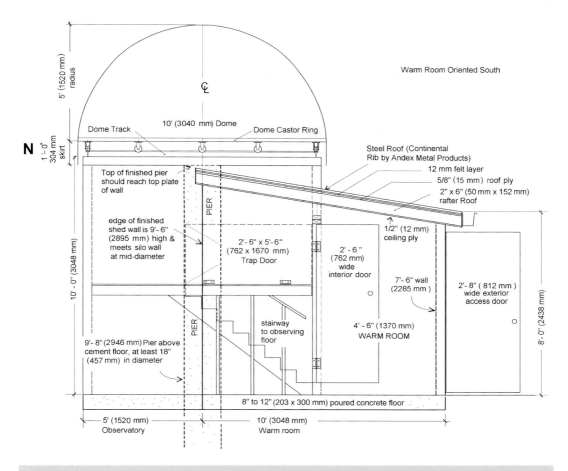

Fig. 13.21 Elevation of a 'warm room' addition showing staircase into the dome, the interior door, and the 4'–6" (1370 mm) wide working space exterior to the cylindrical walls of the observatory [design and drawing by John Hicks]

For most instruments and standing astronomers, this is a little too high to allow a horizontal view of the horizon, and a small rolling ladder will be a necessity for access to near-horizon viewing (I observed horizontally in my 7 ft high-wall observatory quite easily, with a 5 ft high rolling ladder). The warm room design has several other benefits. The roof provides an easy access to the dome exterior for sealing, repairing and painting. As pointed out earlier, access to a free-standing exterior dome without a warm room (or any attached building), is very risky with a ladder that easily slides off its round surface. The addition of a warm room offers a more weather-proof main entry door located well away from the dome. In a free-standing dome, ice and snow can slide off its hemispherical surface, freezing in an icy mass at the door, preventing entry. In areas of moderate to high snowfall, the warm room roof will accumulate snow, but if it is covered in steel roofing, the snow will slide off naturally as temperatures rise, and accumulate outside of the drip-line of the soffit. The decision to build a warm room may require a building permit that could be an unwelcome hindrance to your freedom of construction or locating an observatory. Many detailed plans in the Roll-Off-Roof

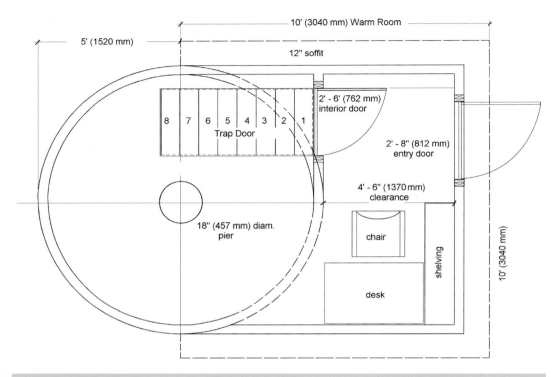

Fig. 13.22 Plan view of the 'warm room' addition showing the staircase with trap door swung up, the interior door, and proposed arrangement of working space possible [design and drawing by John Hicks]

Observatory section of this book contain information that will assist you in the construction of a warm room addition should you wish to add it. These are: Figs. 3.2, 3.4, 4.9, 5.2, 5.4, 5.17, 6.1, 6.2, 6.4, 6.6, 6.8–6.11, 9.1, 11.1, and 11.3.

The design shown in Figs. 13.21 and 13.22 displays the smallest footprint that satisfies the important qualities of ceiling height, usable floor space, and practicable stair-well to create a functional warm room.

Overall Site Requirements

Orientation of the Observatory

A multitude of factors determine how the observatory should be orientated. For solar observers, an observatory with a warm room should be placed with the warm room pointing north to eliminate any solar heating of the steel roof during the day. The rising hot air column emanating from the warm steel roof will most certainly influence 'seeing'.

For regular night-time astronomy, the warm room can be orientated south, as the steel roof loses its heat much quicker than a shingle or composite roof, and it will have little effect at darkness on seeing (see Chap. 2 "Ground Surfaces, Turbulence and Light Pollution"). As pointed out earlier, in areas of moderate to high snowfall, if the warm room is oriented south, its steel roof will naturally shed snow as temperatures rise. Wind velocity on higher elevations such as mountain slopes can affect the slot door operation if not sheltered, with ice or snow jamming its operation from time to time. Most likely the dome section will be placed outward for maximum visibility, leaving little chance to clear snow or ice from its surface. An observatory facing a tree-line, should be placed as far away as possible from the obstruction to increase the angle of sight at the horizon. This becomes increasingly important if the view toward the tree-line is southward from the observatory.

Location and Access

The access to the observatory also influences its orientation. If the foregoing conditions are met, the observatory can be placed for ease of access to a path or roadway as long as the priority for a southern view is retained (a northern view in areas below the equator). For convenience in loading equipment into the observatory, a warm room entry door is best placed convenient to the access route, but for a free-standing dome observatory, this is a little less critical. However, I have always oriented the entry doors on any of the free-standing observatories that I designed on the south side, to benefit from solar heating that melted ice and snow away from the door. On closing up for the day (or night) I also rotated the dome to lock with the slot door facing south for the same reason. Make very sure you can access the observatory for any repairs or alterations that may be required in the future (such as pouring a cement floor for a warm room you might wish to add). If you are lucky, it won't interfere with other structures (such as a garage, tile bed, or power line that already exists). Be very aware of a septic tile bed that is close to your dome observatory. Placing the observatory too close can result in future problems such as odour, or the requirement for tile repairs in the future. Digging up a tile bed is a messy excavation, and if it is too close to your observatory a contractor may force you to move the observatory—a disastrous event. Trees are a concern, but usually careful pruning of unwanted limbs can relieve you of their interference—particularly for a sight line to Polaris, the North Star. Southward and westward views are of most concern since trees anywhere in that line of view must be removed.

Zoning and By-Law Limitations

Refer to Chap. 2 under "Zoning: The Surrounding Land Use" and "Zoning Limitations on Your Own Property" for all the factors involved with city zoning and by-law restrictions that may also be applied against your locating a dome observatory. Similar zoning restrictions apply in larger, rural properties, but they are more accommodating. The same site areas (side-yard, rear-yard etc.) are under regulation in rural areas, but much larger in dimension due to more open space. The dome with warm room attached will likely be required to comply with the existing by-law since the added square feet will surpass the 100 sq. ft (10 sq. m) exception. As a result, you will likely be required to place the warm-room dome observatory at the limit of the rear yard set-back shown in Figs. 13.23 and 13.24. This is precisely why most dome observatory builders choose the free-standing dome—*for exemption from the by-law*. Considering all the factors involved with locating the dome observatory, the builder must choose a path capitalizing on the viewing opportunities. Make a list of factors influencing your site and choose carefully. To neglect these factors could be a costly mistake.

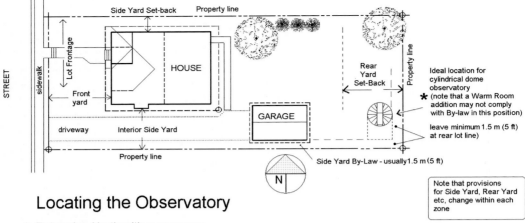

Locating the Observatory

1. Determine North with a compass
2. Find out the minimum sideyard and rear yard restrictions (Town By-Laws) and measure from your lot lines inward (string a line between survey bars)
3. Position the observatory building to take advantage of south, southwest and westward viewing opportunities
4. Make sure you can sight the North Star from the selected position
★5. Note that if you have a warm room attached, the total area is over 100 sq. ft. (9 sq. m) which forces you to conform to the by-law for standard buildings. This will require you to conform to the rear yard set-back

Fig. 13.23 Zoning by-law restrictions limiting the location of the observatory on a typical lot [diagram by John Hicks]

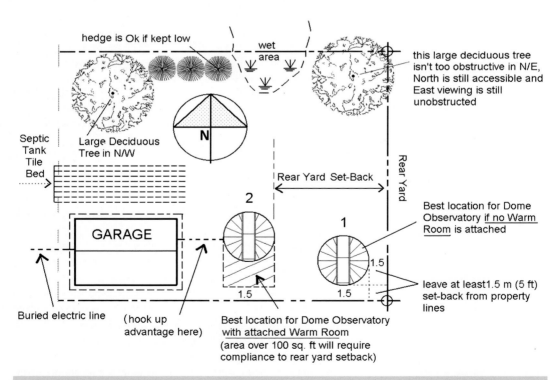

Fig. 13.24 Enlargement of Fig. 13.23 showing the rear yard with various physical constraints combined with normal by-law restrictions [diagram by John Hicks]

Obtaining a Building Permit

In the case where your dome observatory base diameter is over 10 ft, you are likely going to need a building permit. If you are a good draughtsman, you can adapt the plans shown in this book for your purpose, by expanding the diagrams and adding your specific measurements. You will need to use an architect's scale rule and a calculator, or the use of a good drawing program such as Corel Draw if you draw it with a computer. For some, who can translate pixels to inches or mm, you can use the simple Paint Program in your Windows Accessories folder to draw perfectly acceptable diagrams—its not easy but it is possible. Paint is my main drawing program, used in drawing the bit maps exhibited throughout this book. Your plans will have to include the detailed elevation drawings, cut-away details, and a site plan similar to those illustrated in this book.

Many of the drawings I have illustrated will suffice as they are, being only explanatory in nature, while others must be to scale and carry the exact measurements of your model. The Site plan will take the form of Fig. 13.23 with dimensions outlining exactly where it is placed on your property. If you buy a commercial model over 100 sq. ft (9 sq. m) you will likely only have to do a site plan and a foundation plan, the complete observatory already built in kit form or sections, ready to bolt together. The foundation plan must include your footing types (such as sono-tubes) or the cement floor that the observatory is to be placed on. A town engineer may scrutinize your plans offering advice on alternative solutions to design, and he/she will make sure that in approving your details, the town will suffer no liability from lack of safety or structural flaws in design. The security of the dome in high winds, or in rotation and retention on the track, will be high on his/her list of concerns. A major concern will be due to the fact that most 'roofs' are secured strongly to the building under-neath, whereas a dome simply 'sits' on its track, Safety issues involving the possible disconnection of the dome from the track will be high on the list of any plan review, so you must provide details on tie-downs or brackets showing how they will keep it on the track, occupied or unoccupied. Many astronomers build first and take the penalty (if caught) for lack of a building permit. I suggest you keep within your area building regulations, talk to the building department and zoning officer first, seek their advice, and build to code. After all, an observatory is usually permanent, long-lasting, and expensive, and not something to take a risk on. Later on, if you wish to expand (i.e. into a larger warm room) you can apply for a zoning exemption or minor variance. If your application is refused, you still have a standing observatory—an existing, permitted structure. If you submit too large a design in a single submission, you could lose the whole endeavor, due to zoning restrictions. You stand a better chance of success in being granted permission through a minor variance later, since your obser-vatory is both educational and scientific in nature, deserving merit.

The following site requirements are discussed in further detail in Chap. 2 under the heading "Overall Site Requirements", and apply equally to dome observatories:

Polaris and The Southern Sky
Access
Electrical Service
Elevation and Seeing
Soils and Drainage Suitability for Footings
Ground Surfaces, Turbulence and Light Pollution
Zoning: The Surrounding Land Use
Zoning: Limitations on Your Own Property
Ownership Versus Leasing

Chapter 14

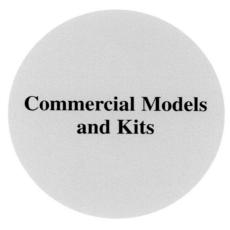

Commercial Models and Kits

The three main observatory designs presently in demand by amateur astronomers are:

1. Roll-Off Roof Observatory.
2. Rotating Clamshell Observatory.
3. Rotating Aperture Slot Observatory.

All three models serve as ideal storage with varying degrees of protection for the observer and equipment. The second and third dome observatory designs include an observing slot or a clam-shell section, that when opened, provides the observer the best protection from cold winds and stray light.

Three commercial models stand out as being the most worthy for amateur astronomers in my opinion—the SkyShed POD, the new-designed SkyShed POD MAX, and the Pulsar Observatory. These are very affordable, unique systems, manufactured in molded/materials, engineered to reduce cost and assembly problems, while providing ease of operation and durability.

The SkyShed POD and POD MAX

Manufactured in high density double-walled polyethylene, the SkyShed POD measures almost 8 ft in diameter (2438 mm), and 8 ft in height. It has had an enormous impact on the astronomical community. This observatory is offered in a range of colors with an option for a luminous 'glow-in-the-dark' model. The luminous dome is an eye-catcher in any location, and serves as a visible facility in an isolated situation where it can be easily found. Departing from the usual fixed hemispheric dome, the SkyShed POD dome is fabricated in two hemispheric half-domes (or clam-shells). However, the mechanical operation differs from that of a 'clam' in this case where one clam-shell is fixed vertically while the other rotates upwards to nestle inside of it. Both half-domes are free to rotate on the base whether the dome is in the closed or open condition.

© Springer-Verlag New York 2016
J.S. Hicks, *Building a Roll-Off Roof or Dome Observatory*, The Patrick Moore
Practical Astronomy Series, DOI 10.1007/978-1-4939-3011-1_14

Fig. 14.1 A standard SkyShed POD Observatory with closed clam shells, built on a circular cement slab footing [courtesy of SkyShed Observatories]

Since both clams rotate together on the horizontal axis, this arrangement permits observation of the sky at all points of the compass right up to the zenith. The SkyShed POD also offers a unique extra shutter unit that swings down containing the familiar open slot found in regular dome observatories. This shutter acts as a combined observing slot and light shield when you need it, normally nestled within with the two clam-shells, and rotating along with them to face any point in the sky (Figs. 14.1, 14.2, 14.3, and 14.4).

The new SkyShed POD MAX is a typical rotating hemispherical dome with a built-in observing slot, and has the benefit of more privacy, light-shielding and wind protection. The SkyShed POD MAX dome is larger in diameter at 12 ft 6″ (3810 mm) with the usual structural framework under a regular dome, and an open slot section with slot door.

Both the SkyShed POD and the SkyShed POD MAX offer additional 'bays' that protrude from their observatory walls to provide space for a small computer station, laptop, LCD monitor and observing accessories. Such work spaces are normally absent from standard dome observatories, and are a welcome addition formed into the regular polyethylene walls of both the SkyShed POD and SkyShed POD MAX observatories (Figs. 14.5, 14.6, and 14.7).

The SkyShed POD dome can be purchased alone minus its base, and can be mounted on a base of your own construction. For some builders, this option allows them to match the exterior cladding or brick work etc., of their home or garage. It also enables the astronomer to house a long focus instrument with a high spatial requirement, or to install a full height entry door. Purchasing just the SkyShed POD dome requires matching the top plate of your self-constructed base to the exact circumference of the dome, allowing for the skirt overhang all around. It also requires you to install castors or rollers of your own choice on the base portion—wheels or rollers pointing upward to run on the underside of the clam-shell dome sections. A unique system well-suited for this purpose

Fig. 14.2 A standard SkyShed POD Observatory with its clam swung up and telescope exposed [courtesy of SkyShed Observatories]

Fig. 14.3 A view of the upward swinging clam nestled inside the fixed clam, showing the rotation point and the ball transfer rollers allowing rotation of the dome [photo by Brian Colville, Maple Ridge Observatory, Ontario]

involves installing 16 or so 1 in. (25 mm) diameter ball transfer rollers forming a smoothly operating "castor ring" under the clam shell half-domes. These may have to be elevated slightly with wooden blocks to keep the clam-shells clear of the base portion of the observatory (Fig. 14.8). Brian Colville's unique ball transfer roller system is shown in Figs. 14.3 and 14.4.

Fig. 14.4 Brian Colville exhibiting his telescope system inside the SkyShed POD. Note the mounted ball transfer rollers supporting the dome [photo by Brian Colville, Maple Ridge Observatory, Ontario]

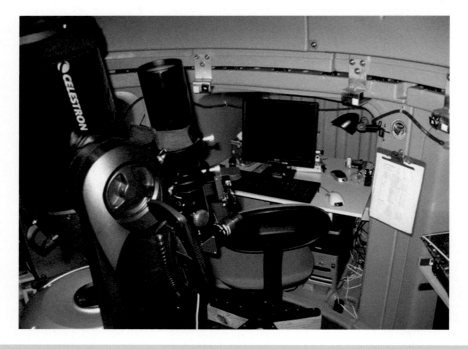

Fig. 14.5 The interior of a well-equipped SkyShed POD showing a small desk and computer station inside a bay [courtesy of SkyShed Observatories]

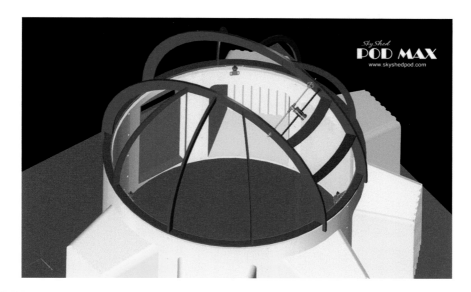

Fig. 14.6 An illustration of the new SkyShed POD MAX showing its dome framework resembling a true observatory dome. This observatory will operate like a traditional dome observatory [courtesy of SkyShed Observatories]

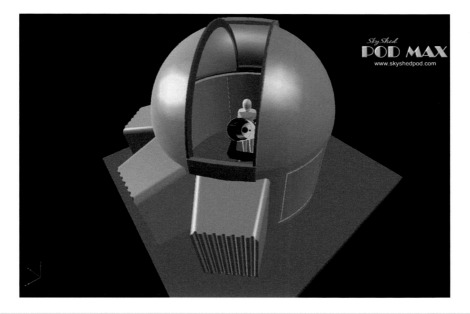

Fig. 14.7 An illustration of the new SkyShed POD MAX showing the completed dome with its slot door section open as in a traditional dome observatory [courtesy of SkyShed Observatories]

In addition to the upward-pointing dome castors (rollers), several 'dome-retaining' castors must be mounted just inside the clam-shell domes to run on a track, fastened inside the edge of the base (usually five castors). A narrow aluminum angle or channel strip rolled to fit the inside radius of the base, and screwed firmly to it, serves as a suitable track. This arrangement keeps the dome from wandering

Fig. 14.8 Brian Colville's finished SkyShed POD on his home-built base, matching the exterior of his existing roll-off observatory [photo by Brian Colville, Maple Ridge Observatory]

on its supporting rollers and from sliding off the base as it is rotated. One of the 'dome-retaining' castors should be spring-loaded to provide both compression against the track and enough flexure to prevent the other castors from binding (Figs. 14.9 and 14.10).

The 'castors' that come with the complete SkyShed POD observatory are typical roller-blade type wheels imbedded into the polyethylene castor ring of the base section, pointing upward. The 'dome-retention' castors on the SkyShed POD package are constructed with an extended steel lip that hooks under the castor ring preventing removal of the clam-shell domes by wind, etc.

Features of the SkyShed POD Observatory

- strong, high density, non-toxic, UV resistant, double-walled polyethylene construction
- lightweight, individual panels
- short assembly time (with only ten fitted panels)
- huge viewing window, horizon to horizon, east to west, 360° rotation
- smooth, easy rotation
- large enough to accommodate a 14 in. Schmidt-Cassegrain telescope
- can be installed on the ground, on a deck, or on a platform (any flat surface)
- wind-tested
- theft prevention through a locking door, and anchor points around the base
- an assembly DVD making assembly easy
- an owner photo gallery to share tips on assembly or equipping the Pod

Fig. 14.9 Brian Colville's dome retention roller system consisting of a grooved pulley running on a rolled aluminum angle. Note the compression spring on the bracket he designed allowing both tension and flexure [photo by Brian Colville, Maple Ridge Observatory, Ontario]

Fig. 14.10 The new SkyShed POD Visor pulls down from the nestled clam-shell domes to provide a familiar observing slot to keep out wind and unwanted light. Note the "PZT shelf on the right hand side of the dome that allows the clam shells to be pushed off their track permitting a view past the Zenith [courtesy of SkyShed Observatories]

SkyShed POD Accessories

The SkyShed POD has an impressive list of accessories to make observing easier or more comfortable:

- optional equipment Pod bays (up to 5)
- black-lined Pod bay option for better dark adaption at computer, etc.
- insulated wall panels 'Ecomate'
- flow-through ventilation air transfer system
- SkyShed POD is available in a wide choice of colors plus a 'glow-in-the-dark' Option
- a 'PZT'—30 in. wide (760 mm) shelf that attaches to the top of the base—allowing the clam-shell domes to be pushed of their track onto it, creating a full clear view past the zenith

The Pulsar Observatory and Technical Innovations 'Home Dome' Observatories

Somewhat similar in design, these two sources produce a traditional, time-tested design erected in countries around the world. The rotating aperture slot dome design has been the choice of most astro-imagers for decades. Pulsar Observatories, located in the UK has distributors in the USA, and Technical Innovations headquarters are located in Orlando, Florida with various distributors in the USA.

The Pulsar Observatory

Pulsar offers 2.2 m (7 ft) and 2.7 m (8 ft to 10 in.) dome models, with a choice of full height and short wall models. Pulsar lists the following attributes to consider in choosing their observatory (Figs. 14.11 and 14.12):

- excellent wind protection for both astronomer and telescope
- minimum exposure to stray light
- excellent dark adaption
- minimal dew problems
- less cool-down time
- fully adaptable for automation
- provides severe weather resistance
- long-life durability of glass-reinforced plastic
- lightweight yet easy-to-rotate dome
- maintenance free
- easy to dismantle and move

The low mass design and air-flow characteristics provide optimum thermal equilibrium for the telescope and its optics, as well as vital protection from high winds, dew and frost. The advantages to the observer are equally beneficial in regard to wind and dampness protection, as well as the convenience of almost immediate employment/deployment of the telescope. Security is high with a full-height observatory that locks from the outside with a key. The dome slot door locks securely from the inside rendering the observatory completely tamper-proof from the outside. Pulsar observatory domes are recognized by a broad base of demanding institutional and amateur users, offered at a fairly modest price.

2.7 METRE FULL HEIGHT OBSERVATORY DIMENSIONS

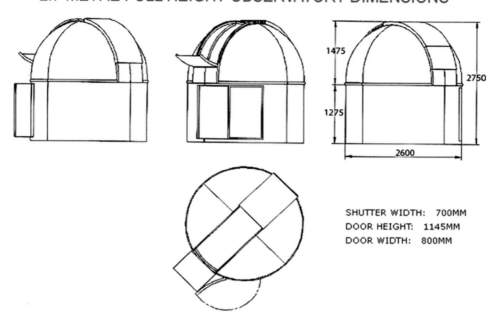

SHUTTER WIDTH: 700MM
DOOR HEIGHT: 1145MM
DOOR WIDTH: 800MM

Fig. 14.11 Diagram illustrating the Pulsar 2.7 m (8 ft to 10 in.) full-height observatory with significant dimensions [courtesy of Pulsar Observatories]

Fig. 14.12 View of Pulsar 2.2 m (7 ft) dome observatory with slot door shutter closed sitting on a cement pad [courtesy of Pulsar Observatories]

Motorized Rotation Drive

Pulsar offers a fully-motorized, rotation drive system powered by a 12 V DC lead-acid battery. Using a solar-panel charger supplied by Pulsar, the unit can be trickle-charged during daylight hours. Dome slewing at 400 times sidereal speed is possible in either direction by remote operation , using a key fob transmitter. This has a range up to 300 ft (100 m). A microprocessor controls the automatic tracking by moving the dome in a stepwise fashion, each step being approximately 1.8°. Tracking speed can be adjusted from ten times sidereal down to sidereal, and as low as 1.8° of rotation. The motorized dome drive is a must for unattended long-exposure photography and imaging. Slewing the observatory dome in line with the telescope allows your telescope and dome to simultaneously track a chosen object, without your moving the dome (Figs. 14.13 and 14.14).

Motorized Slot Door Shutter Drive

A motorized slot door drive system, also powered by a 12 V DC lead-acid battery, is available to open and close the slot door. The slot door is controlled by a three-position switch with the central position being off. The shutter is controlled by moving the three-position switch to the open or close positions. The kit is supplied with battery, controller, cables, motor, and necessary hardware.

Manual Slot Door Pulley System

For those who do not require a powered slot door drive system, the pulley accessory kit is designed to aid in the opening and closing of the dome slot door. It minimizes the risk of the shutter slipping and causing damage, and also allowing the shutter to be part-opened in any position. Easy to fit, the kit comes with all equipment and 13 ft (4 m) of nylon rope, supplied with detailed fitting instructions (Fig. 14.15).

Fig. 14.13 View of Pulsar Rotation Drive system showing the on-board controller [courtesy of Pulsar Observatories]

Fig. 14.14 View of 2.2 m (7 ft) Pulsar dome with slot door shutter open and attached solar panel charger [courtesy of Pulsar Observatories]

Fig. 14.15 View of the Pulley Accessory kit offered in place of the Motorized Slot Door Drive System [courtesy of Pulsar Observatories]

Observatory Bays

Pulsar will supply full-height observatory bays to store computers, LCD monitors, and accessories for digital astro imaging. It is advisable to purchase bays at the time you purchase your observatory, ensuring that the wall panels are pre-cut to receive the bays at no extra cost, saving a later delivery charge. The accessory bays come with a fixing kit and full installation instructions (Fig. 14.16).

Standard Observatory Piers

Through the research of Astro Engineering, Pulsar has been able to offer an observatory pier that uses AE's anti-vibration design technology. Manufactured from heavy steel, the Standard Pier is finished in black stipple-finish resin powder coating. Several types of CNC machined heads are available to fit popular German equatorial mounts. Two pier models are available specifically designed with a side aperture (or "Owl's nest") to accept EQ5 and EQ6 Sky Watcher mounts. These piers feature the high precision CNC machined head to provide ultra smooth and precise adjustment for accurate polar alignment and the most demanding CCD imaging (Figs. 14.17 and 14.18). There are also several other pier models available:

- Standard pier for altaz & wedge (LX200, LX90 etc.)
- Pro-Pier for equatorial wedges (LX200 etc.)
- Do-it-yourself pier kit for equatorial wedges (LX200 etc.)—the kit provides all the parts and hardware needed (you make the concrete pier)

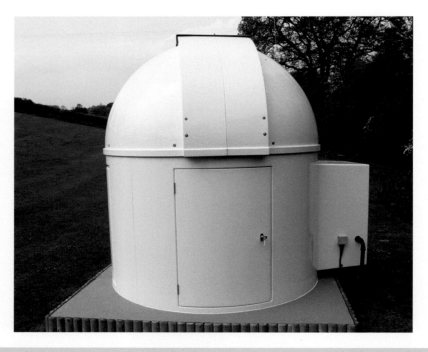

Fig. 14.16 View of Pulsar observatory with attached bay and slot door shutter closed [courtesy of Pulsar Observatories]

Fig. 14.17 View of Pulsar Standard Steel Pier manufactured from heavy steel and finished in black stipple-finish resin powder coating [courtesy of Pulsar observatories]

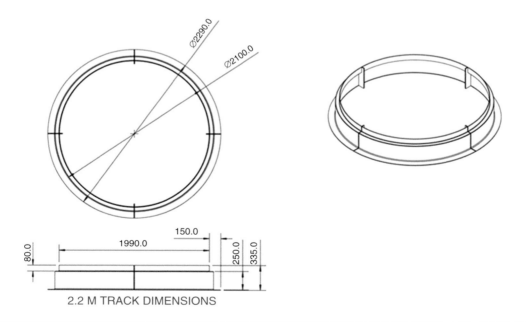

2.2 M TRACK DIMENSIONS

Fig. 14.18 Diagram illustrating a complete Track for the Pulsar 2.2 m (7 ft) model observatory dome offered separately for those clients who want to make their own base or marry the dome to a roof or other suitable structure [courtesy of Pulsar Observatories]

- Pro-Pier for Losmandy GM11 and GM8, with CNC marry plate
- a 2.2 m (7 ft) complete Track unit is available, separate from the base, for those clients who want to make their own base, or for those who wish to fit it on a roof or other suitable structure

  Several other items complete the Pulsar Observatory accessories list:

- Impact resistant interlocking rubber tiles for the observatory floor, that are shock-resistant, providing insulation from a concrete base, reducing vibration and noise
- observatory alarm system deterring intruders (has built-in PIR detector and telephone dialer)
- wireless camera link (used in conjunction with the rotation drive allowing remote operation of the observatory dome by providing a visual link from the observatory to the warm room or computer room)
- dome security clamps providing additional security for the observatory—preventing the dome from being removed by any external force (wind storm etc.)
- all-metal observatory eyepiece shelf (fits into rear of the dome top and rotates with the dome)

Technical Innovations 'Home Dome Observatories'

The Technical Innovations Observatory is designed primarily for stand—alone installations. The popular 10 ft (3 m) Pro Dome model (PD 10) is a fiberglass observatory that can accommodate Newtonian telescopes up to 16″ diameter and Schmidt Cassegrain telescopes up to 16″ diameter. The Pro-dome model has a molded semi-door section to allow full height entrance, additional rollers and soffit to meet the most extreme wind and snow conditions. It has a key-operated, custom dead-bolt lock for security. The package offered includes dome quadrants, rear cover, shutters, dome support ring, base ring (with or without the molded semi-door section), special locking hardware and fittings, silicone sealer, with instructions for construction and operation.

You can add motors to rotate the dome, move the shutters, and even operate the dome remotely. All hardware comes with stainless steel bolts.

Dome Construction

The dome is made of two pairs of quadrants which bolt together along internal flanges, and sit on a 12″ high base ring using hard rubber ball bearing rollers for dome rotation. The finished dome turns easily on the hard rubber rollers mounted on the base ring. A reverse flange above the base ring covers the roller area and retains the dome from shifting in wind. The Pro Dome uses bolt-together construction, and assembly requires alignment of parts, measuring and drilling bolt holes, and the use of common hand and power tools. Larger holes, for rollers and latches, are cut out and factory-finished. Typically, domes are assembled in place by two or more people, without a crane or special equipment. The PD 10 ten foot diameter domes are installed at homes, schools, colleges, museums and research facilities, and can be placed on concrete pads, towers, decks, rooftops, garages etc.

Shutters

The shutter area is a generous 36″ wide extending to the zenith and 16″ beyond. The shutters open by sliding up and over the dome, automatically disengaging during opening, to nest together outside on the back of the dome. Two integral, automatic latches lock the shutter together for security when it is closed.

Observatory Wall

The walls of the observatory are formed from modular 'wall rings'. These can be stacked to match your telescope and pier height. The wall rings can maximize the usable space inside the observatory, granting nearly a 6 ft high clearance right against the wall. The model offers a full height entrance door employing the slot opening as part of the doorway, which allows access without ducking under the dome ring or requiring high walls.

Weather-Sealing

Weather-protection is assured by overlapping flanges and baffling, not employing seals that wear out or deteriorate. Temperature control is also offered by the brilliant white gel exterior, and little cool-down time is required before your observing session begins. A dark blue interior finish helps to preserve dark-adaption. Because no part of the PD 10 Home Dome Model weighs more than 45 lb, it is easy to assemble without the use of a crane (Figs. 14.19 and 14.20).

   Both the SkyShed POD, SkyShed POD MAX, Pulsar Observatory, and Technical Innovations Home Dome are well-suited for amateur astronomers who either lack the skill to assemble a self-built model, or lack the time to do so. The observatory design that follows is a replica of my observatory, self- built and designed by myself. It outlines in detail the construction of a free-standing observatory (that is, without a warm-room), since it is the model most-often sought after by amateur astronomers who are limited sometimes by space and zoning regulations.

Fig. 14.19 The Technical Innovations PD 10, Pro Dome on a concrete slab

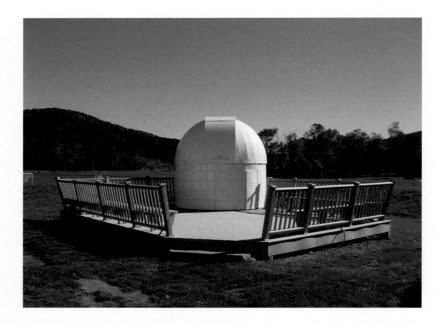

Fig. 14.20 The Technical Innovations PD 10, Pro Dome on a raised deck

Chapter 15

Plans and Construction Steps Toward Building a Dome Observatory

Introduction

Before you begin construction, consult both Chaps. 2 and 13 sections on "Overall Site Requirements" and "Site Suitability", to make sure you have selected the best site for your dome observatory. (See Figs. 2.5, 13.23, and 13.24).

Nothing is worse than having chosen a poor site when a better location was available. The following 12 major steps are sequential in nature, allowing you to proceed from the ground up. If you find yourself stuck on a particular assembly, go back and review these steps and figures. Each one is important in such a complex construction. Essential to building this observatory is access to several fabricators and parts supply shops. The dome construction, for example, requires access to a roller bending mill. The accurate bending of the strong aluminum angle cannot be over-emphasized, for in such a spherical structure, the framework elements must fit precisely. Also a helio-arc welder is required for joining the sections of rolled angle together to complete the basic dome skeleton. You can skip visiting the roller bending mill by constructing the arches, ribs, slot door and its frame in laminated plywood (much like the sole plate and top plate of the observatory base wall).

From experience with other wood framework domes, the application of the aluminum panels is easier over wood members, but eventually the wooden framework rots from either leakage or condensation against the inside surface of the aluminum panels, and the whole construction is compromised. The all-aluminum dome model, although more difficult to construct, will last a lifetime. My dome, now 35 years old looks like it was built yesterday.

With respect to welding the aluminum framework, I suggest that you do **not** helio-arc weld the two arch-ends that connect to the dome base ring, nor the ends of the ribs that connect to both arch and base ring. The reason for this is that if bolted, they can be adjusted to create a continuous accurate width throughout the slot door opening, assuring that the slot door does not bind. Once welded, anywhere, this option is lost. The pre-rolled sections of the aluminum dome base ring should be welded to form a perfect circle, and the slot door braces welded between the arches, once the slot door gap is assured accurate.

© Springer-Verlag New York 2016
J.S. Hicks, *Building a Roll-Off Roof or Dome Observatory*, The Patrick Moore Practical Astronomy Series, DOI 10.1007/978-1-4939-3011-1_15

Dome Observatory Construction Stages—12 steps plus Maintenance
Basic Observatory parts list
Fabricating the Metal Pier
Site preparation, pouring the concrete base, pier, and the floor slab
Framing the cylindrical base, walls, and pre-hung entry door.
Fabricating the Track and V-Groove caster assembly
Preparing the Dome Framework, roller bending and welding requirements
Assembly of the Dome framework and applying the aluminum panels
Construction of the Slot door frame, slot door, and its wheels
Construction of the elevated observing floor
Lifting the Dome onto the track, securing the dome, and adding the skirt
Installation of mechanical aids, doors, locks, dome rotation device etc.
Weatherproofing the structure
Final luxuries, and interior fixtures
Maintenance

Basic Observatory Parts List

The following list categorizes the major parts to build the observatory from the ground up, and serves
as a check list as you proceed through construction.

Foundation cement, forms, rebars and anchor bolts
Sono-tube Pier, cement, rebars, piping, and Steel Cap
Aluminum Pier Top, and end-plates
Electrical wire, PVC tubing, Electrical boxes, electrical conduit
Walls and Base ring parts (Base Ring Ply, Stud Wall, Upper Plate Ply Ring)
Pre-hung steel Door and double bolt locks
Track Assembly (Fabrication and Welding)
V-Groove Castors (5 minimum)
Dome Ply Ring
Dome rotation mechanism (Castor, pillow blocks, chain drive or ship's wheel drive)
Aluminum Dome framework parts (Dome Base Ring Sections, Dome Arches, Front-side Dome Slot
Door Guides, Slot Door Frames, Back-side Dome Slot Door Guides, Slot Door top End, Ribs, Back-
side Slot Braces)
0.025" aluminum sheets (dome skin)
Fastener parts (Stainless bolts, nuts, washers)
Interior wall light (red light)
Infra-red alarm system (security)
Wall Fan Exhaust with weatherproof outside grill
Exterior Vinyl Board and Batten siding
Interior and exterior Masonite sheeting
Stainless cable (Dome Tie-down mechanism)
Aluminum finishing strips
Pier rug wrap, and Floor rug
Coatings and finishes
Sealants and caulking materials

Fabricating the Metal Pier

Early in the construction process you must decide how you want to position the pier within the area of the floor slab. It can be either offset from center or in the center of the floor slab. Schmidt-Cassegrain telescopes, short in nature, will place the observing eyepiece in a position close to the pier, whereas refractor and Newtonian style instruments, generally much longer, require considerable swing around the pier.

If the southern sky is mainly your viewing objective (it usually offers the majority of deep sky objects) you can offset the pier southward giving you more space for yourself positioned in the northern sector of the observing floor. This works well for Schmidt-Cassegrain instruments but not long-focus refractors, and Newtonians. If pointed toward the northern celestial sky they will cramp the observer against the southern wall of the observatory. If your long-range plans for telescope models are not clear, its best to locate the pier dead-center.

The next preliminary task is to calculate the height of the pier relative to the finished floor level. If you are building the high-wall dome observatory featured in this book you will need a two-piece pier (concrete + metal pier). Calculate the vertical height of your telescope mount plus telescope pointed in a horizontal position, and subtract this from the height of the castor ring (the upper ring holding the dome and the castors is called the "castor ring"). In the model illustrated in this book, it measures 8 ft from the floor. The difference will give you the overall height of the required pier. Split this in half to determine the two lengths of concrete and metal pier sections required above the observatory floor. Add four feet to the lower, concrete section, for the required buried footing below the floor, creating a total concrete pier length of approx. 7 ft long. (Refer to Fig. 4.1, for details on this calculation). Figure 15.1 illustrates the construction of the upper metal pier manufactured from 10 in. diameter thick-wall aluminum tube.

It should be at least a 3/16″ thick-walled tube approx. 3′–6″ long (depending on your mount-plus-telescope height). Figure 15.2 shows the finished aluminum pier section with its end caps. Make sure you measure the bolt circle on the concrete pier below it accurately, and pre-drill the bolt holes carefully in the upper pier cap before you assemble the aluminum pier. Use flat-heat stainless bolts to fasten the end caps to the tube, counter-sink the heads and tap the holes for $\frac{1}{4} \times 20$ thread with a hand tap.

After finishing construction of the upper pier store it away for the time being as it could be easily damaged by the continuing construction. Note that you could helio-arc-weld the caps onto the tube if you prefer to save the extra machining, thread-tapping etc., but make sure the bolt circle in the top cap fits the holes in your mount, and the bolt circle in the bottom cap fits the holes in the concrete pier cap—before its welded on.

I do not recommend fabricating a full length concrete pier for a high wall dome observatory. Not only is it extremely difficult to concrete-fill such a high sono-tube form, it will be structurally unsuitable. Because of its extra length, a pier made in two different materials (concrete and metal) will null any resonance and vibration faster than a pier composed of only one material. In case you decide to terminate the observatory wall at 4 ft, building a low-wall model, you can use a short concrete pier terminating at the concrete floor level and bolt on an alternative steel or aluminum pier shown in Fig. 4.6. This pier also contains a push-pull arrangement on its top plate, which allows you to level your mount and telescope. Figure 4.7 illustrates a detail for construction of the end caps drilled to match both the top of the concrete pier and your telescope pier. You can also employ a total length of concrete pier for a low wall observatory with approximately 3 ft above the floor, and four ft below the floor and into the ground, as pointed out earlier. This will still be structurally stable, reducing your costs (see Fig. 4.1).

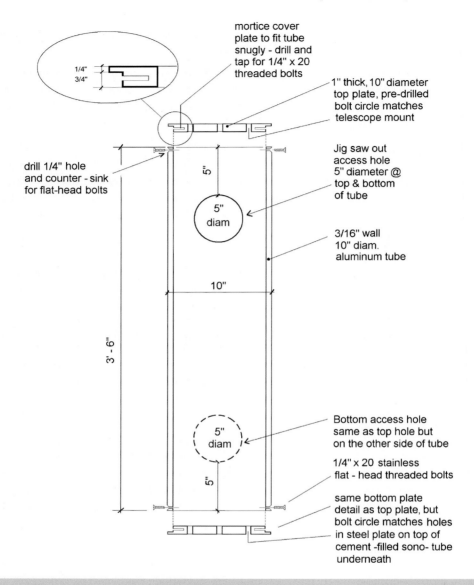

mortice cover
plate to fit tube
snugly - drill and
tap for 1/4" x 20
threaded bolts

1/4"
3/4"

1" thick, 10" diameter
top plate, pre-drilled
bolt circle matches
telescope mount

drill 1/4" hole
and counter - sink
for flat-head bolts

5"

Jig saw out
access hole
5" diameter @
top & bottom
of tube

5"
diam

3/16" wall
10" diam.
aluminum tube

10"

3' - 6"

5"
diam

Bottom access hole
same as top hole but
on the other side of tube

1/4" x 20 stainless
flat - head threaded bolts

5"

same bottom plate
detail as top plate, but
bolt circle matches holes
in steel plate on top of
cement -filled sono- tube
underneath

Fig. 15.1 Detail of upper metal pier construction

Finally, a steel cap must be made for the concrete pier top for either high-wall or low-wall observatory, pre-drilled to accept three 18″ long × ½ in. threaded bolts that will be plunged into the wet concrete-filled pier. The upper aluminum pier unit will have its lower plate drilled to accommodate this bolt circle. The cap should be made from minimum ¾″ thick steel and the same 10″ diameter as the concrete pier. Secure the bolts initially with nuts on both sides of the cap, removing the top ones once the concrete has set. You should have this plate made, and ready, <u>before you pour cement into the pier</u>. Refer to Fig. 4.8 and read the section "Full length concrete pier" for advice on construction.

Fig. 15.2 Photo of upper metal pier with its end caps

Site Preparation, Pouring the Concrete Base, Pier and Floor Slab

Pouring the Pier

Begin by marking out the floor slab circumference using a fixed center point (such as a dowel driven into the ground) and a 5 ft length of string. Mark out the circle on the ground, keeping the string taught as you sweep around the circumference.

Excavate all the earth within this area to a depth of 4 in. When this is level and smoothed start to excavate the ground for your pier with a post-hole digger or a long-handle shovel to a depth of 4 ft minimum with a diameter larger than the sono-tube form chosen for your pier. Place the excavated soil in a wheel-barrow and dump it <u>outside</u> the floor slab area, retaining the last wheel-barrow full for back-filling the hole around the form. Calculate the length of electrical conduit and wire required to reach from the inside wall (about 3 ft up the wall), across the radius of the floor (5 ft) and up the pier at least 5 ft, plus some extra wire for the electrical boxes. Allow for a total of 15 ft of wire and PVC conduit to contain it. Slide in the sono-tube form, plumb it with a carpenter's level, and begin back-filling it to a point where it is free-standing.

You will then need to brace the sono-tube in position with lumber, angled off to the outside of the floor circle because the concrete load to be poured into it will easily topple or break-off the upper portion of the sono-tube. Beware of this and brace it well. Years ago, I poured a pier with a friend, and we had forgotten to brace it. Amounting to hundreds of pounds of wet concrete, it began to lean to one side, requiring one of us to support the tube until it could be braced by the other. The last step before pouring concrete into the pier is to fashion a rebar cage down the inside of the sono-tube forming a crude triangle with coat-hanger wire holding the cage together. Use three 5/8″ rebar strands 18″ shorter than the length of your cement pier (they should be about 5.5 ft long). Lower this cage into the tube letting it rest at the bottom of the tube, and brace it internally from touching the inside walls of the sono-tube. Before you begin mixing the concrete, make sure the sono-tube pier is plumb and rigid, and also that you have the steel pier cap ready with its threaded bolts attached. It is a good idea at this time to add the electrical boxes (I suggest you use 2 four-outlet boxes to the outside of the sono-tube, one at the bottom and one near the top). Make sure you position them on the side where your electrical conduit and wire locate. Drill through the sono-tube wall using the electrical box bolt holes as a template. Cut the heads off long stainless steel ¼″ × 20 threaded bolts and insert them through the sono-tube wall protruding into the soon-to-be-filled concrete interior, leaving enough thread outside to hold the electrical boxes.

Then add the electrical boxes and fasten with stainless nuts. (The reason for this is that you will have to remove the boxes later to insert the PVC conduit and wire). Tape both boxes solidly against the sono-tube form and bag with plastic to keep any spilled concrete off them. Attend to these boxes and their position as you pour the concrete into the sono-tube, being careful not to knock them out of position (see Fig. 4.9, for detail "wiring a Grounded Duplex receptacle").

Add water slowly to the premix bagged concrete in an old wheel barrow, and when it reaches the consistency of oatmeal porridge scoop it out with a bucket and pour it into the sono-tube. Use a long pole to pack the concrete uniformly to eliminate air pockets.

This step is very important as air pockets will weaken the pier. Take your time in pouring and packing the concrete. When the tube is full to the top let it sit awhile to settle completely. Carefully plunge the bolted metal pier cap (long bolts down) into the concrete, wiping away any excess. This may take some effort as the concrete will resist your efforts. (Now you can see why it was essential to leave 18″ off the length of the rebar cage to accommodate the bolt lengths.) When the cap is flush with the top of the sono-tube, use a level to make sure it is completely level. If it is not, carefully knock the cap with a block of wood and hammer until it is. Do not commence pouring the floor slab on the same day and let the pier dry thoroughly to avoid disturbing the braces (Fig. 15.3). (Refer back to Fig. 4.8 for a photo of the finished pier.)

Construct a batter-board circular form with 3/8″ plywood (not chip-board) all around the perimeter of the 10 ft diameter floor. Use 2″ × 2″ stakes to support the plywood form on 2 ft. centers (the wet cement will push hard against this with considerable force). The stakes go on the outside of the circle, driven at least 1 ft into the ground. Nail the plywood to this from the inside, and check the accuracy of the radius of each stake-plus-plywood. The plywood form top should be 8 in. above the existing leveled grade (make sure the existing grade is level all around before you begin setting the forms). Place the measured conduit <u>with wire in it</u> on the excavated floor extending from the pier to the perimeter. Add PVC <u>rounded</u> 90° elbows to each end, passing the electrical wire through them, gluing them with PVC cement. The outside elbow (at the plywood form), should be in position such that it will pierce the plywood sole plate when it is finally placed on top of the finished cement floor—about 2 in. inside the forms.

At least 3 ft of wire should be left protruding from the 90° elbow on the perimeter, and also 3 ft protruding from the 90° elbow at the pier (depending on where you position the lower-most electrical box). At the pier, add straight conduit to fit the lower electrical box, pulling the wire inside the box. You will have to detach the box to do this, then glue the PVC conduit into the box and re-attach it with the nuts, flush to the sono-tube surface. Wrap the excess wire up in a coil and tape it also to the pier securely.

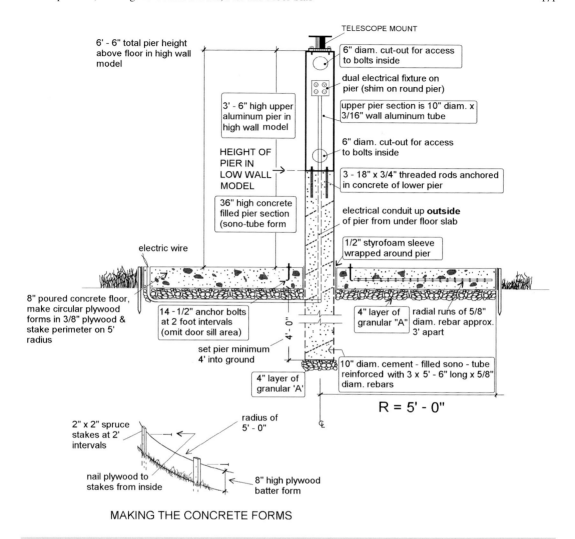

TELESCOPE MOUNT

6' - 6" total pier height above floor in high wall model

6" diam. cut-out for access to bolts inside

dual electrical fixture on pier (shim on round pier)

3' - 6" high upper aluminum pier in high wall model

upper pier section is 10" diam. x 3/16" wall aluminum tube

HEIGHT OF PIER IN LOW WALL MODEL

6" diam. cut-out for access to bolts inside

3 - 18" x 3/4" threaded rods anchored in concrete of lower pier

36" high concrete filled pier section (sono-tube form

electrical conduit up **outside** of pier from under floor slab

electric wire

1/2" styrofoam sleeve wrapped around pier

8" poured concrete floor, make circular plywood forms in 3/8" plywood & stake perimeter on 5' radius

14 - 1/2" anchor bolts at 2 foot intervals (omit door sill area)

4' - 0"

4" layer of granular "A"

radial runs of 5/8" diam. rebar approx. 3' apart

set pier minimum 4' into ground

10" diam. cement - filled sono - tube reinforced with 3 x 5' - 6" long x 5/8" diam. rebars

4" layer of granular 'A'

R = 5' - 0"

radius of 5' - 0"

2" x 2" spruce stakes at 2' intervals

nail plywood to stakes from inside

8" high plywood batter form

MAKING THE CONCRETE FORMS

Fig. 15.3 Cross-section of the concrete pier and floor slab. This figure shows the entire assembly of both concrete pier and floor in one diagram with measurements and details (the upper aluminum pier extension is shown but not attached at this time)

If you are constructing the 7 ft high wall observatory, the <u>upper</u> electrical box is added later, when you add the upper aluminum pier section. If you decide to build the 4 ft low-wall observatory, position the electrical box high up on the sono-tube close to where the metal pier cap and your telescope mount will be situated. If you plan to use an all-metal pier, you can coil the wire in a bag on top of the finished floor-level pier where its metal pier cap will sit waiting for the prefabricated metal pier. (See Fig. 4.6. Detail showing construction of a typical metal pier.)

The other end, at the outside circumference of the floor, should be inserted in a short length of conduit, about 10 in. long—just enough to pierce a future hole in the plywood sole plate of the wall. Glue this to the 90° elbow with wire in it and coil the wire in a plastic bag on the existing outside soil surface to keep it clean from the cement pour. (See Fig. 4.4 for a drawing of a typical slab with electrical service.)

To separate the floor from the pier, you need to wrap a ½ in × 12 in. band of styrofoam (preferably ethafoam) around the base of the sono-tube pier such that it completely surrounds the circumference of the pier (no gaps), then tape the ends together. You are now ready to add the 4 in. layer of granular A crushed gravel base.

Fill the entire excavation to a depth of 4 in. with crushed granular "A" gravel, and rake it smooth and level.

Pouring the Floor

Ideally, you will need a portable cement mixer to fill the floor since it will require quite a quantity of concrete. If you can get at it with a cement truck—its so much easier, and no more expensive than mixing it yourself with pre-mixed concrete. Arrange a ramp to wheel barrow the concrete from truck or portable mixer to the site. Mix the concrete until it reaches the consistency of porridge and fill the floor circle to a depth of 4 in. At this point rake it relatively smooth and lay 5/8″ rebars on the concrete in a radial pattern about 3 ft apart at the circumference, terminating before they touch the pier and its styrofoam wrapping. Fill the rest of the floor with another 4 in. of concrete on top of the rebar to create an 8 in. depth of concrete floor. Level this and smooth it out with a long 'screed' board fashioned from a 5 ft length of 1 in. × 6 in. pine.

Before the concrete sets add the ½″ diameter anchor bolts ("J-bolts") at 2 ft intervals around the perimeter about 2-½″ in from the edge of the concrete, exposing about 3 in. of threaded bolt above the concrete surface. Do not place an anchor bolt in the section where the entry door is to be located.

Keep the concrete lightly damp with a very fine spray of water every hour or so until nightfall, to prevent the floor from drying too fast and cracking (Fig. 15.4). (See Fig. 5.5 for a detail of how "j-bolts" [anchor bolts] fasten down a sole plate.)

Framing the Cylindrical Base, Walls, and Pre-hung Entry Door

The fabrication of the circular plywood rings (sole plate, top plate, and dome caster ring) require jig-sawing out circular arcs of plywood, and laminating them in three layers to form the cylindrical base rings and supporting ring of the dome. These circular rings must be both glued and screwed together to form rigid structural components. If you make these yourself you will need an hand-held jig saw and saw-horses to hold the material. They are cut out from sheets of 4′ × 8′ × ½″ plywood in lengths that will fit the sheet. Depending on how long you decide to make these sections, the plywood can be most economically utilized by the manner in which you fit them on the sheet—obviously the longer arcs (up to 8 ft) will take up much of the sheet because of their arc. I decided early to job this function out to a large wood-working shop to allow me to carry on with the dome construction. The bottom ring (sole plate) and the top ring (top plate) are both 5 ft in outside radius and 4′–7″ inside radius creating a width of 5″.

The Dome ring (Caster ring) has a <u>larger</u> outside radius of 5′–2″ and an inside radius of 4′–7″ (same as the lower two rings)—the extra 2 in. of width outside diameter creates an overhang for the aluminum skirt which is secured to the dome. This makes the Dome diameter a total of 10′–4″. Each of the three rings are a 3-layer lamination of ½″ ply, glued, screwed together, and stacked so that the seams do not coincide—to provide more strength. After assembly, each ring will measure approx. 1-¾″ thick due to the glue layer, and plywood surface roughness (the flathead screws should lie flat with the surfaces). The process requires many sheets of ½″ plywood and is laborious. If you do it yourself you will need a garage floor or large workroom to accommodate the task. My late friend

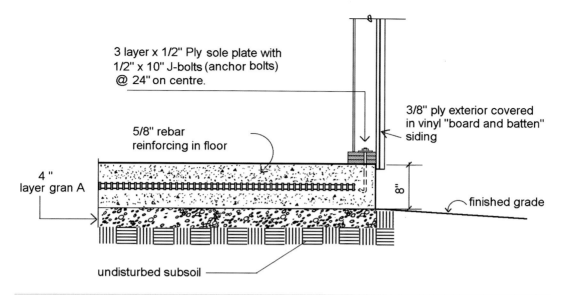

Fig. 15.4 Detail showing the position of anchor bolts (J-bolts) and the rebar in the concrete floor. Note that the anchor bolts pass through the plywood sole plate in the middle point between the studs

Walter Wrightman scribed the ring radii out on his workroom floor and stacked the plywood arcs accurately upon them as he assembled, glued, and screwed the sectors together. Once assembled, leave the rings flat on the floor until you are ready to assemble the base. Paint these rings with at least two coats of spar varnish at this time to weatherproof the concealed areas. When dry, take the base ring and top ring over to the observatory site and place them on blocks—get help to do this so they do not bend or de-laminate, because they will flex considerably.

Place the base ring carefully over the anchor bolts, align it plumb with the concrete perimeter (use a straight edge) and rap the plywood ring sharply to indent it where the anchor bolts lie underneath. Then remove the ring, turn it over and drill ¾″ diameter holes right through the 3-ply layer for the anchor bolts (if you are unlucky to hit a screw on the way through you will have to cut through it). Remember to mark where the conduit and wire will pass through from the slab and drill a hole to accommodate the short conduit tube. It is wise at this point to place arcs of asphalt-impregnated board underlay continuously lining the area where the base ring (sole plate) will seat. This will require careful fitting so that no gaps exist—you can trim any overhang later on with a sharp box cutter. This step will save you the annoyance of ants and other insects getting in under the lip of the exterior panel and eating the sole plate.

Place the drilled sole plate in the right orientation over the anchor bolts, pass the electrical conduit and wire through the sole plate and secure the plate with nuts and washers (Fig. 15.5).

Begin toe-nailing the pre-cut 2″×4″×7′ long wall studs into the base ring flush with the concrete base with two ardox galvanized nails per stud. After four studs are placed at corners of a compass, someone has to help you hold the upper top plate into position over the studs, and butt-nail them through the plywood ring. A step ladder is required—to attain the 7 ft high level. That done, the remaining studs can be placed 16″ on center all around the perimeter with someone butt-nailing the tops into the top plate while another toe-nails the bottoms into the bottom plate.

Make sure to leave a rough opening for the pre-hung steel door (the door instructions will tell you how much to allow), but stick to the 16″ centers on studs, with the last two on either side of the door frame being an odd-sized gap (make a note of this gap). Prior to this step you will have purchased a

Fig. 15.5 Photo showing how the base ring is secured with anchor bolts

pre-hung plain steel door—no window, and with no embossing. It should measure 7′–6″ high and 2′–8″ wide,—this size door fits the 5′ radius best. Next, you need to brace the wall studs 4 ft up their length with curved segments. To do this, cut curved sections out of 2″ × 10″ pine board that will offer an outside radius of 5′ with an inside radius of 4′–8-½″, providing a width of 3-½″.

Make up a thin plywood or thick cardboard template to mark out the curved section needed. Cut these at least 2″ longer than your stud gaps (longer than 16″) because you will have to cut them off to fit them snugly between the studs. Hold the braces against the studs to scribe the cut-line parallel to the flanks of the 2″ × 4″ studs. It is slow work, but important, because these braces also serve as nailing points for both interior and exterior masonite sheathing.

If adjacent braces are slightly offset in an alternating manner, they can be butt-nailed through the 2″ × 4″ studs providing a much stronger grip than toe-nailing them (Fig. 15.6).

To frame the pre-hung steel door, refer to Figs. 6.6 through 6.11, for complete instructions. The door sill plate (at the bottom of the door) will be straight, while the door top header will be curved— set about 4″ down from the plywood top ring.

Cut this from a 2″ × 10″ pine board (as with the wall braces). This header is important as it maintains the curvature of the wall and the 2″ gap between it and the dome skirt overhang from the dome. Pull the wire up through the conduit that now pierces the base ring and lead it up to a single electrical duplex receptacle about 1′–6″ above the sole plate ring. Wire it into this receptacle according to Fig. 4.9, cut off the excess wire, and screw the box with its face protruding about 1/8″ beyond the inner face of the stud.

Check the base completely, making sure nothing is missing such as another wall fixture you may want, or another electrical receptacle to serve a computer station etc. When you are satisfied, apply 1/8″ sheathing over both outside and inside walls, but pre-paint them with spar varnish before applying them allowing them to dry thoroughly. If you are planning for only a solar observatory, you can use pre-coated white masonite panels on the inside creating a bright interior (Fig. 15.7).

Fig. 15.6 Photo showing how the circular wall braces fit the curve of the base ring, staggering up and down between the wall studs for butt-nailing

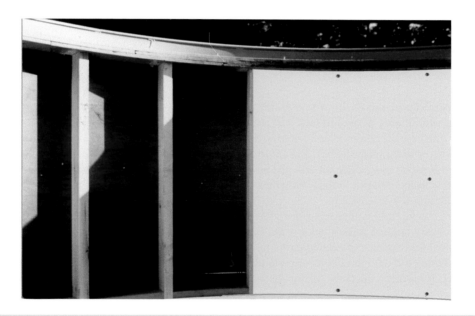

Fig. 15.7 Apply the pre-coated white masonite sheets forming the inside wall surface for a solar observatory, or paint raw, uncoated masonite sheets with flat black paint for a deep sky observatory

If you are planning for a deep sky observatory, you should paint the inside flat black to prevent reflections. I fastened the interior panels with $1/8'' \times 1''$ aluminum strips over the masonite seams with round head stainless screws and S/S finishing washers, which allowed me to remove panels for future wiring etc. Use flat head aluminum nails to fasten the <u>outside</u> masonite panels. Install the outside siding of your choice over the masonite, which is easier to install now, than later, once the skirt is added to the dome. I like vinyl board and batten siding in a cream color which matches well

Fig. 15.8 Finished application of vinyl board and batten exterior siding. Note the black rubber inner tube ring at the base of the vinyl slats that I used to water-seal the base ring from moisture. This rubber strip, installed under the vinyl and lapping over the concrete lasted several years but eventually failed requiring annual caulking around the entire observatory wall perimeter. Make very sure that you lap the siding down over the concrete base least 2 in.

with the shiny aluminum dome. If you prefer, for maximum protection, you can wrap the outside in 0.30 guage aluminum, sheet sold in $4' \times 8'$ sheets. This will require drilling holes in the position of studs and braces to accommodate stainless steel screws (use only stainless or a bi-metal reaction will corrode the aluminum sheets). You must also caulk the seams with silicone, and the aluminum application is much more difficult and just as expensive as the vinyl siding. Make sure, at all costs, that you extend the exterior siding—whether vinyl or aluminum, over the edge and down the concrete foundation for at least 2 in.

Failure to do this will compromise the waterproofing of your observatory. My exterior vinyl siding was erroneously applied to butt against the finished concreter floor (which was slightly more than 5 ft in radius) leaving an exposed concrete lip which has resulted in water leaking in under the base ring and running out onto the floor. No matter how well I caulked this bottom seam it leaked in heavy rain storms. It has been the curse of my observatory—so learn from my greatest mistake (Fig. 15.8).

Fabricating the Track and V-Groove Caster Assembly

The Steel Track

A ¼ in. thick × 2″ tall steel bar must be rolled in a roller mill to a radius of 4′–9-½″ and welded together to form the track circle. Like many of the other structural components required, the track was rolled in 8 ft sections, to be welded together in a metal fabrication shop (find a handy one for you will be visiting it periodically).

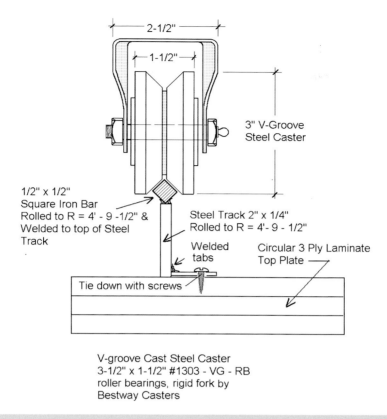

2-1/2"

1-1/2"

3" V-Groove
Steel Caster

1/2" x 1/2"
Square Iron Bar
Rolled to R = 4' - 9 -1/2" &
Welded to top of Steel
Track

Steel Track 2" x 1/4"
Rolled to R = 4'- 9 - 1/2"

Welded
tabs

Circular 3 Ply Laminate
Top Plate

Tie down with screws

V-groove Cast Steel Caster
3-1/2" x 1-1/2" #1303 - VG - RB
roller bearings, rigid fork by
Bestway Casters

Fig. 15.9 Cross-section detail of the track, with the V-groove caster sitting in place on top of the knife edge iron bar, showing the track tab screwed down to the base top plate

You must specify both the exact radius (4′–9-½″) to be rolled to, and the diameter (4′–9-½″ × 2 = 9′–7″ diameter). In order to maintain a rigid, transportable, and accurate circle, the shop will have to make a "spoked-jig" to hold it together. This is achieved by welding short tabs to the inner face of the track every 2 ft that bolt to spokes which run to the central point of the circular track. Iron bars ½″ × ½″ will suffice, drilled at their outer ends to bolt to the tabs, and welded firmly together at their terminus in the center of the track circle. (The spokes are unbolted one by one as you screw down the track, and that is why they are simply bolted to the tabs.)

Finally, the shop must roll an iron bar ½″ × ½″ to the same radius as the steel track and weld it (spot-welds) every foot or so on either side to the top of the rolled track bar. This forms the running surface for the V-groove steel rollers that will be secured under the Dome Base Ring (called the Caster Ring) (Fig. 15.9).

Before delivery of the "spoked-jig" and circular track, you must check the radius all a around its circle with a measuring tape held precisely over the knife edges across its diameter. Do so before it is delivered, or before you pick it up. Note that it measures as an oversize load so it must be tilted and shipped on edge to avoid an escort. Mine was braced with chains across a ½ ton truck bed. The track should measure no-more than 1/8″ off 9′–7″ diameter, but if it does you can correct this by pushing or pulling the segment in or out before you screw it down. It won't matter much if it's a low amount off round because you can correct it as you install it., but if its too large an error it will not fit on the base top plate, and you would have to cut it and re-weld it.

Fig. 15.10 The track "spoke-jig" positioned on the top plate with one bolt removed from a spoke, the tab ready to screw down

The important measure is the ultimate screwed-down radius, because if its not done accurately, the casters will ride up on the beveled iron bar and bind, making rotation difficult. The process involves placing the "spoke-jig" on top of the base top plate <u>centering it all around,</u> and screwing the tabs down as you move around the circle, checking the radius as you go. Undo each tab from its spoke, one at a time, and screw it down—but not fully. If slightly out-of-round push or pull the segment to achieve the 4′–9-½″ radius. Of course, the original shop circumference must be correct or no amount of push-pull will create a perfect track. You must make the shop aware of this (Fig. 15.10).

Begin by unbolting one spoke (any spoke) and screw the tab down to the plywood top plate. When through screwing down all the tabs check it for centering on the top plate You can then lift off the "spoke – jig" and discard it (take it to a scrap dealer). Finally, level the track. If it's not level in places undo the tab screws in that area and slide 1/8″ or ¼″ shim strips (about 2 in. wide) under the track until level. When you are assured, tighten down all the tab screws firmly. Spray the entire track and Base top plate with two coats of Tremclad or other similar rust-proofing product. It is essential to do this now as you cannot get at the caster gap easily later, when the dome is over it.

The V-Groove Caster Assembly

Refer back to Fig. 15.9 for details about the steel casters. You will need six of these 3-½″ × 1-½″ number 1303-VG-RB with roller bearing centers and grease nipples.

Bestway Casters manufacture these as catalogue number 1303-VG-RB or select similar, but make sure they are V-groove, same diameter, width, and height, made of steel, and designed to rest on a ½″ × ½″ steel edge (Figs. 15.11 and 15.12).

Fig. 15.11 Casters shown attached to the dome base ring (caster ring), perfectly aligned on the steel knife edge of the track. You should be able to rotate the ring with one finger, with the minimum friction afforded by the V-groove casters

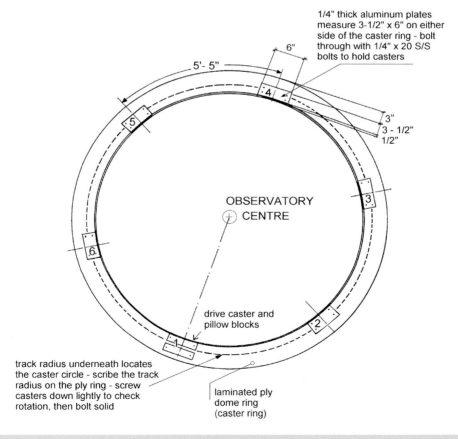

Fig. 15.12 Illustrates the location of the casters, bolted to the dome base ring. They situate on 5′–5″ centers (measured on the circumference) around the underside of the dome base ring, bolted through the base ring

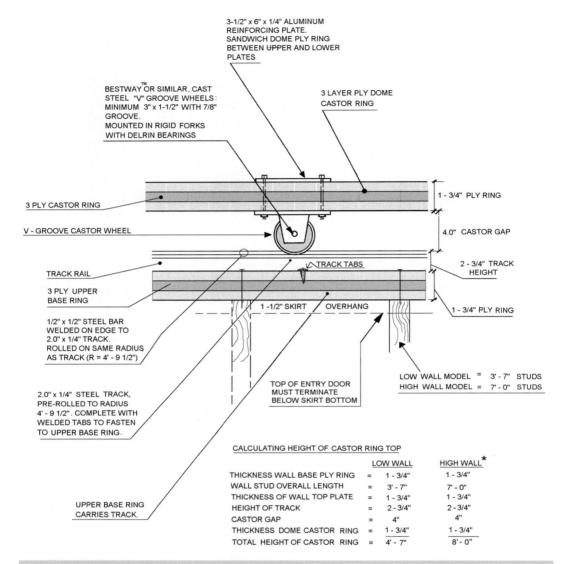

Fig. 15.13 Diagram illustrating the final castor gap between the top of the steel track and the underside of the dome base ring (caster ring)

The caster assembly is reinforced both on the upper side and lower side of the laminated ring with aluminum plates, the bolts passing through them. These should be ¼″ thick and 3-½″ × 6″ in size, secured with ¼″ × 20 threaded stainless steel hex-head bolts, from caster plate right through the laminated ply ring and the upper plate. These guarantee no future shift in the position of the casters that might occur in time through torque passing through just the wood laminate.

They also serve as lightening connectors from the aluminum dome down the casters, into the track, and through a heavy gauge copper wire to a ground outside and down the wall of the observatory. These are extremely important plates—do not eliminate this detail. Later on in the assembly of the dome, we will install connecting strips of aluminum to these plates from the aluminum dome frame ring guaranteeing a good conductivity to the ground.

The resulting gap between the track and the dome base ring (caster ring) should be about 4 in. as shown in Fig. 15.13.

Fig. 15.14 The interior rug skirt is pulled back to reveal a caster on the track, the laminated dome base ring (caster ring) and the top plate of the observatory wall

   This is the "air gap" that will allow air circulation in your observatory, and it will be covered by a rug "skirt" inside to eliminate dust, insects, and wind-driven rain from getting inside the observatory. Note that there is a pressure differential between the inside and outside of the dome during a high wind because wind driven over the dome travels further than wind driven straight through the caster gap-creating lift. This airfoil phenomena can cause your dome to lift off in high winds. The interior rug and exterior aluminum skirts help somewhat, but you must tie-down the dome on leaving it—a detail to be shown later on (Figs. 15.14 and 15.15).

Dome Drive Mechanism

One of the casters of the six will be a drive caster, engineered to turn the dome. It is not necessary to motorize this mechanism, although some will attempt to. It is an easy task to rotate the dome manually every 20 min or so no matter what type of viewing or photography you are undertaking, and a manual mechanism is foolproof, needing no electrical connection. This is best achieved by placing the sixth caster wheel between two roller bearing pillow blocks, removing it from its normal caster forks and running a 1″ diameter steel shaft through its central bearing hole. Purchase two pillow blocks with grease nipples and 1″ central bearing hole plus a 1″ diameter steel shaft about a foot long to fit. Remove the forks from the caster, and the ¼″×20 grease nipple. Insert a ¼×20 allen-key grub screw into the former grease nipple thread. Position the two pillow blocks on the underside of the laminated dome ring such that they sandwich the caster directly over the track, leaving a "wiggle" space of about 1/8″ between the caster and both pillow blocks so the caster won't bind on irregularities in the track. Tighten down the allen key screw on the steel shaft leaving a little projection only on the outside, but 6–8″ on the inside for attaching a chain wheel or boat wheel. Cut a wood spacer to fit under the pillow blocks such that the caster will just rest on the track. You will likely need two pieces of wood ¾–1″ thick, and a some trimming to arrive at the required thickness.

Fig. 15.15 The effects of a wind storm on a dome observatory in Arizona—the roof fully disengaged and instrument damage. [Courtesy of Arizona Sky Village]

When satisfied with a good fit, remove the pillow blocks, drill bolt holes through the wood to match those in the blocks, and bolt the assembly under the dome laminated ring, right through, with washers and nuts on top. Lightly bolt the pillow blocks at first and rotate the caster ring on the track to make sure it runs freely. When satisfied it will rotate well, the next step requires some adjustments.

The drive caster must push down on the track slightly more than any other caster, and this is achieved by inserting a ¼″ thick rubber plate under each pillow block (yes, you will have to remove the pillow blocks again to insert it). When you attach a rug skirt on the inside later on, you will have to punch a hole through it for the 1″ shaft. You can either purchase a chain wheel about a foot in diameter to fit the 1″ diam. bearing or find a small ship's wheel to rotate the dome. I prefer a ship's wheel because I can rotate it easily and faster. It is also not as cold to turn as a chain loop in colder weather (Figs. 15.16 and 15.17).

Other Track Options

Several other possibilities for track construction are possible. Some builders have used large roller bearings running in a channel, but most end up with hard rubber casters that run on a wood surface with lateral casters to maintain the dome on the observatory. Most of the rubber or hard composite caster products either flatten through sitting so long idle, or add too much friction running on wood. Ball transfer rollers, or skate board casters work much better providing you employ a plastic or metal running surface for them (usually the rollers or casters are positioned on the base pointing upward with the running surface attached under the Dome laminated ring—the idea being that the running surface remains dust and dirt-free). Golf balls have been used in a half-pipe rolled to the required

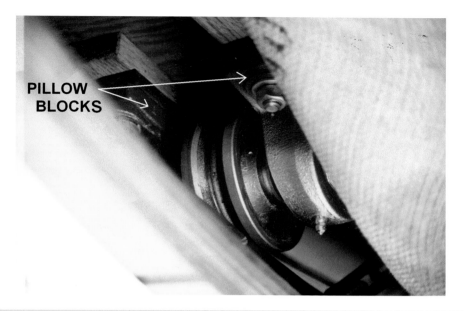

Fig. 15.16 Photo of the pillow blocks sandwiching the caster wheel on the underside of the dome ring (caster ring). The interior rug skirt has been pulled back to show the full apparatus

Fig. 15.17 Photo showing a chain pulley rotation device, with its continuous looped chain. The pulley is grooved to grip the chain, and the ends of the chain are joined together by a chain-link after calculating out how far down you want to grip it (depends on your height). We will refer back to this later under "Installation of Mechanical Aids—Step 10"

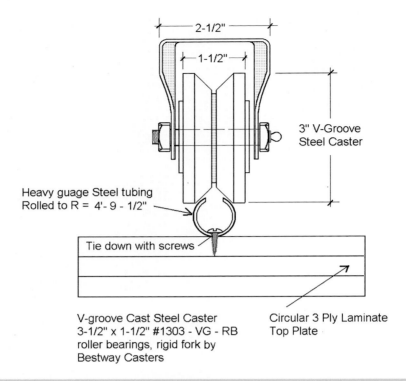

2-1/2"

1-1/2"

3" V-Groove
Steel Caster

Heavy guage Steel tubing
Rolled to R = 4'- 9 - 1/2"

Tie down with screws

V-groove Cast Steel Caster   Circular 3 Ply Laminate
3-1/2" x 1-1/2" #1303 - VG - RB   Top Plate
roller bearings, rigid fork by
Bestway Casters

Fig. 15.18 Cut-away view of a V-groove caster resting on a tube track

radius, but they brush and jamb together, bumping along the running surface due to their dimpled surface. A lateral caster is still needed to keep the dome from slipping off the base. Although the knife-edge track is hard to construct it solves two major problems:

1. it reduces friction to a minimum
2. it eliminates lateral casters

My suggested alternative to a knife-edge track, if some builders find it too difficult to fabricate, is the use of a heavy gauge steel tube rolled to the same radius (R=4′–9-½″). The critical dimension with this type of track is its diameter—too much diameter and the caster can possibly ride off, too little diameter and friction is increased as the tube bites into narrower angle of the caster (the edges of the notch at the apex of the caster groove) (Fig. 15.18).

A possible problem may exist with the screw access hole on the top of the caster, as the caster rolls over it. Casters clambering over large screw access holes will increase friction, and possibly arrest a caster. Two ways to avoid this are (a) choose the smallest diameter screw-heads to reduce the diameter of the screw access hole, or (b) provide a surface for the caster to rest upon at the access hole point (see Fig. 15.19).

Preparing the Dome Framework, Roller Bending and Welding Requirements

Refer back to the "Basic Observatory Parts List" to familiarize yourself with the various structural components of the dome. Purchase the aluminum angle stock required according to the following schedule, and take the members requiring bending to a local roller bending mill.

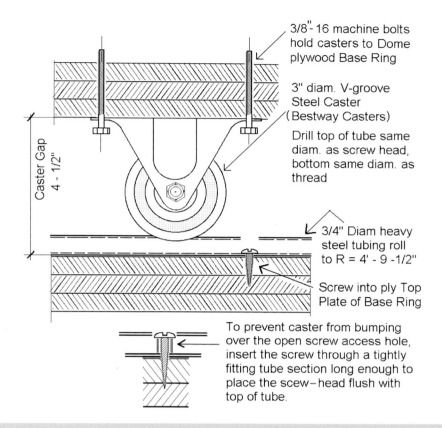

3/8"-16 machine bolts
hold casters to Dome
plywood Base Ring

3" diam. V-groove
Steel Caster
(Bestway Casters)

Drill top of tube same
diam. as screw head,
bottom same diam. as
thread

3/4" Diam heavy
steel tubing roll
to R = 4' - 9 -1/2"

Screw into ply Top
Plate of Base Ring

To prevent caster from bumping
over the open screw access hole,
insert the screw through a tightly
fitting tube section long enough to
place the scew–head flush with
top of tube.

Caster Gap
4 - 1/2"

Fig. 15.19 Side view of an alternative tube track, with a short piece of tubing containing the screw—just long enough to raise the screw head to the top of the track providing a rolling surface for the caster. The screw head may have to be filed smooth and flush with the top of the track tube

Sometimes it is better to have the mill supply your stock because they will choose an aluminum grade that bends easy—check this out first with them. Be very careful to state the required radii and the position of the leg (leg-in or leg-out). If they bend it the wrong way you have wasted that section of angle!

Curved Aluminum Stock required:

4–8 ft sections for Dome Base Ring (2″×2″×3/16″) angle rolled
        R = 5′-2″ Leg-in

4–8 ft sections for Dome Arches (2″×2″×3/16″) angle rolled
        R = 5′-2″ Leg-in

2–10 ft 6″ sections for front side Slot Door Guides (2″×2″×¼″) angle rolled
        R = 5′-2″ Leg-out

2–10 ft 6″ sections for front side Slot Door Frames (2″×1-½″×¼″) angle rolled
        R = 5′-4″ Leg-in

2–8 ft sections for back side dome Slot Door Guides (2″×2″×¼″) angle rolled
        R = 5′-2″ Leg-out

14–8 ft sections for Ribs (1-½″×¼″) bar rolled
        R = 5′-2″

1–3 ft section for upper end of Slot Door Frame (2″×1-½″×¼″) angle rolled
        R = 5′-4″ Leg-in

Linear Aluminum Stock required

4–3 ft linear sections ($2'' \times 2'' \times \frac{1}{4}''$) angle for Slot Braces

1–3 ft linear section ($2'' \times 2'' \times \frac{1}{4}''$) angle for Slot Door Stop

1–3 ft linear section ($2'' \times 1\text{-}\frac{1}{2}'' \times \frac{1}{4}''$) angle for bottom of Hatch

If the mill has a helio-arc welding capability, have them weld the sections of the Dome Base Ring together for you, but remember to include a linear section 27″ long at perfectly opposite sides of the ring for the slot door gap—this is terribly important—that uses up your four sections of Dome base Ring and two of the four linear sections of the Slot Door Braces. Check with the mill first to see if they can deliver it to your site—it is an oversize load (10 ft in diameter) which will otherwise need a police escort. The mill will likely have a way to stand it up on its edge within their vehicle. If you decide to weld all the required members on-site that may be easier, but this will depend on the availability of a portable helio-arc welder. On-site welding is better in that you can get all the dome structural members welded without having to transport an unwieldy structural object along public roads. I emphasize again NOT to weld the ends of the arches to the Dome Base Ring unless your welder is very proficient in lining up the arches perpendicular to the Dome Base Ring (getting an exact 90° angle and clamping it well). If he does not, your slot door width will not be parallel and the slot door will not fit or run true. Its best to simply bolt them to the base ring and adjust their width throughout their arches before welding in the two Slot Door Braces higher up on the dome. After that, if you wish, you can also weld the arch-ends to the Dome Base Ring but it is not necessary. The assembly sequence is outlined next in "Assembly of the Dome Framework".

The four slot door braces (which includes braces 1 and 4 welded into the Dome Base Angle) maintain the flat surface of the slot door section which makes construction of the this area easier. Measure and cut the four slot door braces precisely—they are all 27 in. long and must be welded into the aluminum dome base angle and arches. The top end of the slot door itself will be of curved angle for reasons that will be clear later in the construction process, while the bottom end is linear with no terminal angle.

Assembly of the Dome and Applying the Aluminum Panels

Take the castor ring off the track and place it on wooden blocks (cut offs from railway ties or logs are perfect for this) so that the castors do not touch the ground. Wrap up the castors with plastic bags and twist ties to keep them from getting wet while the dome frame is being assembled.

Construction of the Dome Frame

Get assistance to carry and place the welded-together Dome Base angle onto the top of the laminated ply castor ring. Center it equally all around the plywood castor ring and then bolt it down (use $\frac{1}{4}$ in $\times$ 20 long hex head bolts). The next step is to bolt in the two arches to create the slot gap of exactly 27 in. Make sure the entire arches are uniform in separation (27 in. throughout the slot opening) and clamp it with wood braces. When assured of parallel arches, drill and bolt them into the Dome Base angle using 3/8 in. $\times$ 16 thread size flat head stainless bolts. Use flat head stainless bolts throughout the frame construction and counter-sink flat on the exterior surface so the aluminum skin will lie flat on it (Fig. 15.20).

Once the two arches are rigidly bolted to the dome base ring, the second and third linear braces can be welded in at the positions shown in Figs. 15.21 and 15.22. The slot door stop angle is the fifth linear section used in the dome framework which lies on the exterior of the dome over the finished aluminum panel covering the backside of the slot door section. It will be bolted through the panel and brace number 3 at a later date when the slot door nears completion (Fig. 15.23).

Fig. 15.20 Photo showing Arch junction with the Dome base angle (this junction is shown in the photo as welded which is not necessary)

Adding the 14 Ribs

The next step involves the addition of the 14 ribs spaced equidistant around the dome base angle, curving from the base ring to the arches. These have been rolled previously to a radius of R = 5′–2″ from 1 to ½″ wide × ¼″ thick aluminum bar. Do not attempt to hand-bend these as the material is too thick and the dome will end up non-hemispherical from varying radii (Fig. 15.24).

Before connecting the ribs to the dome base ring and the arches, it is important to create a centroid point at the very apex of the dome between the two arches. Locate this point on a thin strip of 1″ wide × 1′–8″ thick aluminum screwed lightly between the two arches. To do this, measure over the entire curvature of the arches with a tape measure, divide it in half, and install the aluminum strip at this point. Mark the centroid in the middle with thick magic marker (Figs. 15.25 and 15.26).

Each rib is positioned 1′–9″ on center from its neighbor, and bolted to the dome base ring with flat head ¼″ × 20 thread machine bolts. After bolting the rib to the dome base angle, gently position each rib over to its respective arch aiming at the centroid mark. Scribe a line where the curved rib meets the inside leg of the arch, and cut the rib off at this point (you may be required to remove the rib to cut it precisely). Clamp the rib and arch together, drill a hole through both arch and rib, and fasten with ¼″ × 20 thread stainless steel flat head machine bolts. Countersink the flat heads into the exterior arch and dome base angle exterior surfaces. Complete the whole circumference of ribs and then remove the aluminum centroid strip (Figs. 15.27 and 15.28).

I recommend adding an extra rib for strength in the center of the slot door frame, from the dome base ring up through the second linear brace to the third linear brace. This will require piercing the leg on the second linear brace by cutting a slot in it to accommodate the 1-½″ wide × ¼″ rib that passes through it. Do this by drilling 3 or 4 holes in the aluminum leg, filing out the remainder of the slot. The curved aluminum bar should fit easily through it, providing extra strength in this wide area (Fig. 15.29, 15.30, and 15.31).

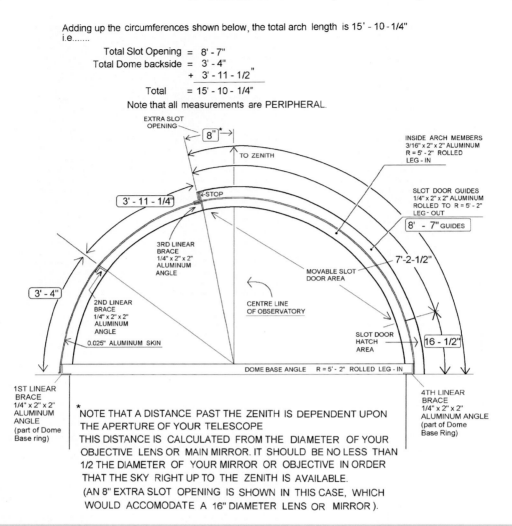

Fig. 15.21 Diagram showing a sectional view of the location of all four braces including the two linear braces (braces 1 and 4) welded into the Dome Base angle. Measurements are shown for the slot opening, brace separations, and arch lengths

Applying the Aluminum Panels

This step can be infuriating unless it is done strictly as outlined. The skirt should be attached first, but cannot be applied until the dome panels are completely installed and the finished dome is raised up on the caster rail. The reason for this is that the crane operator cannot see the castor rail hidden by the skirt and possibly derail the dome.

The principle involved here is that employed in laying shingles (the upper shingle must always overlap the lower not vice-versa, hence the panels must overlap the top edge of the skirt). The skirt should be slightly thicker gauge aluminum (0.030″ thick aluminum is advised). Pop-rivet panels to the Dome angles, arches and ribs with pure aluminum pop rivets (non-ferrous) as shown in Fig. 15.32 "Applying the dome panels and skirt". If a little wrinkling in the aluminum results, the final pinning

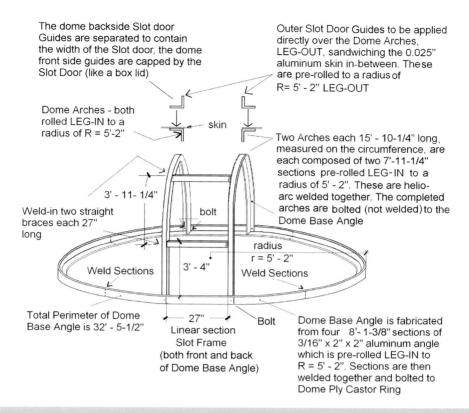

The dome backside Slot door Guides are separated to contain the width of the Slot door, the dome front side guides are capped by the Slot Door (like a box lid)

Outer Slot Door Guides to be applied directly over the Dome Arches, LEG-OUT, sandwiching the 0.025" aluminum skin in-between. These are pre-rolled to a radius of R= 5' - 2" LEG-OUT

Dome Arches - both rolled LEG-IN to a radius of R = 5'-2"

skin

Two Arches each 15' - 10-1/4" long, measured on the circumference, are each composed of two 7'-11-1/4" sections pre-rolled LEG-IN to a radius of 5' - 2". These are helio-arc welded together. The completed arches are bolted (not welded) to the Dome Base Angle

3' - 11- 1/4"

Weld-in two straight braces each 27" long

bolt

radius r = 5' - 2"

Weld Sections

3' - 4"

Weld Sections

Total Perimeter of Dome Base Angle is 32' - 5-1/2"

27"
Linear section
Slot Frame
(both front and back of Dome Base Angle)

Bolt

Dome Base Angle is fabricated from four 8'-1-3/8" sections of 3/16" x 2" x 2" aluminum angle which is pre-rolled LEG-IN to R = 5' - 2". Sections are then welded together and bolted to Dome Ply Castor Ring

Fig. 15.22 Diagram showing the backside of the dome frame with linear braces 2 and 3 in place. This diagram also shows the leg position of the angles used throughout the Dome frame

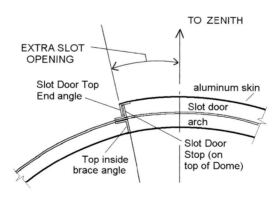

TO ZENITH

EXTRA SLOT OPENING

Slot Door Top End angle

aluminum skin

Slot door

arch

Slot Door Stop (on top of Dome)

Top inside brace angle

Fig. 15.23 Detail of the slot door junction with the slot door stop angle when the slot door is closed. When the door is pulled down tight, this junction is completely weather-sealed at the apex of the dome

of both skirt and panels with a finishing strip will remove them. The dome panel application must be done according to the sequence shown in Fig. 15.32. Rough cut the 0.025″ thick panel to fit the rib gap overlapping on both left and right sides. Starting with the panel next to the slot door section, drill through the panel and arch at mid-section and secure with a 1/8″ rivet to hold it in place. Bend the panel at both top and bottom to make sure it will cover the entire gap, and rivet the opposite rib at

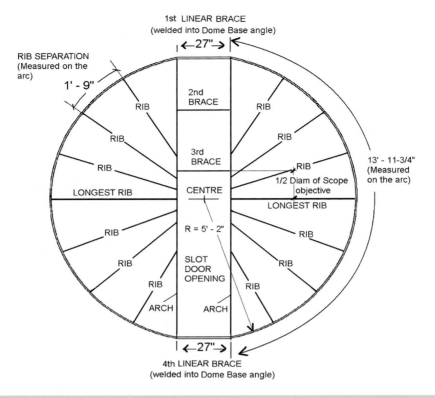

Fig. 15.24 Diagram showing the 14 rib locations spaced 1′–9″ around the dome base angle. Note how the third linear brace is offset from the center (apex) of the dome, creating the extra slot opening past the zenith for vertical viewing. This offset should be equal to ½ the diameter of your expected telescope aperture

Fig. 15.25 Photo of the author's wife, Lorraine, with the aluminum strip to her left showing how the ribs converge toward it

Fig. 15.26 Photo of the slot door opening with the aluminum strip holding the centroid mark at the *top*. Notice how I curved the second and third brace, which made my dome construction very difficult and is not recommended

Fig. 15.27 Bolting the 1-½″ wide × ¼″ thick ribs to the dome base angle. A ratchet wrench and a large screwdriver help here to speed things up. Note the flat head bolt on the outside which must be countersunk to allow the aluminum panel, added later, to fit over it smoothly

Fig. 15.28 The author bolting the last rib to the arches at the top of the dome

Fig. 15.29 An extra rib placed mid-slot door on the back-side of the dome adds extra strength to the structure. Notice how the rib pierces the leg on the second linear brace in order to maintain a smooth surface for the aluminum panel to rest upon

Fig. 15.30 The dome frame nears completion. Notice the extra rib placed in the slot door section to strengthen it

Fig. 15.31 The author's wife, Lorraine and Tanya, standing within the finished dome framework. The slot door opening is clearly visible allowing Lorraine to stand up in the future viewing aperture

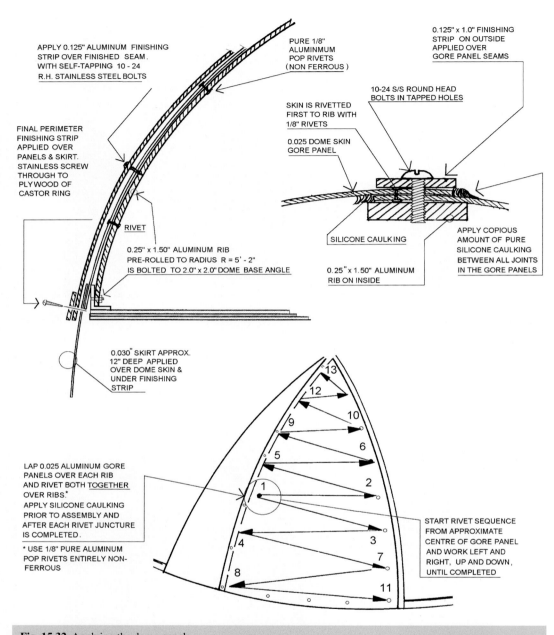

Fig. 15.32 Applying the dome panels

the same height. Moving from left to right, the next aluminum panel is riveted through the last panel creating a double panel layer over the next rib. This means that fewer rivets are needed in the right-hand panel as you proceed, because a double number are added with each panel overlap. Once a rivet holds the panel spread clear silicone sealant prodigiously over the right hand rib before covering it with the panel. This is messy, since the panel will slide around on it as you drill and pop-rivet the panel over the rib, but it seals it underneath.

If you find it too difficult to do—eliminate the sealant application—it can just be used as a caulk along the seams when finished. Proceed pop—riveting both upward and downward alternating from left to right rib locations, continue until you have completed the attachment of the entire panel. If you do not use this riveting sequence, the skin will buckle and pock creating an ugly-looking dome. When satisfied that it is smooth, and evenly applied, trim the panel on both sides to the edges of the ribs and arches, removing all the surplus aluminum. Use the same procedure as outlined for the first panel in riveting sequence, proceeding around the dome, including the backside slot door area, until you arrive again at the front slot door opening. Proceeding around, add a strip of silicone caulking to the seams of each overlapping panel . Be careful to do this very thoroughly as the silicone caulking lasts 10 years if you do it right, and getting up on the dome with a ladder later on to caulk it is very difficult.

When finished with the aluminum panel installation, preferably the same day, install aluminum $0.25'' \times 1.50''$ finishing strips over all the rib-panel seams, drilling and bolting them through the over-lapped panels and the rib underneath. Use stainless round head 10–24 bolts and nuts. If you have to do this alone you will have to tap the holes with 10–24 tap thread, or drill slightly undersize holes to hold the bolts tightly in order that you can put the nuts on inside the dome without the bolts turning. First, apply prodigious amounts of silicone sealant under the finishing strip before bolting it down. The force of the bolts is much greater than pop-rivets, and the finishing strip application squeezes the panels hard against the ribs, eliminating bulges, air spaces etc., and tightens the whole affair increasing its appearance and weatherproofing greatly. The last perimeter finishing strip will go all around the dome, later on in the construction process. It will tie down the dome panels and skirt to the dome ply ring underneath, screwed down with stainless steel screws. The structure should now look like an observatory dome, minus the slot door section in the middle, and the skirt covering the track and caster gap (See Fig. 15.32 for complete assembly).

Choices in Slot Door Design

You may be attracted to the historical dome design with its double shutters that slide apart on tracks to reveal the slot door opening. The two identical halves, both semi-circular, require separate track systems which project beyond the curvature of the dome carrying the doors at their top and bottom edges. These present a much greater challenge to the builder than fabricating a single, curved, roll-over door. There are several advantages to the roll-over (hemispheric) door:

1. it does not present a wind scoop when open, like the exposed halves of the shutter-type which overhang the dome curvature.
2. it is much easier to weather-seal with only a small hinge-out lower section to completely seal the door.
3. it produces much less trouble opening under ice and snow conditions. Snow easily falls through the separating halves of the shutter-type doors as they are opened, entering the observatory—the roll-over door carries it over the backside of the dome.

For these reasons, my choice of design features a hemispheric (roll-over) door, which is shown in the plans that follow. See Fig. 15.33 "Comparison of popular Slot Door Types"

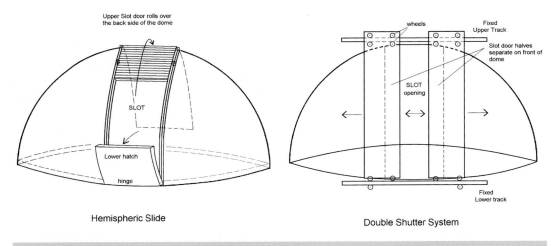

Fig. 15.33 Comparison of popular Slot Door types

Bolting the Slot Door Guides to the Outer Hemisphere of the Dome

The position of the backside and front side (slot side) of the door guides is critical. These twin sets of guides contain the door when it is open, as it slides over the backside of the dome, and also guide the door, as it closes over the observing slot maintaining a weather seal. They must be accurately installed in such a way that they do not impede the movement of the door as it opens or closes, but also seal the closed door from the weather. Figure 15.34 shows how the paired $2'' \times 2''$ aluminum angles switch orientation as soon as the slot door passes the Slot Door Stop sitting at the top of the dome. In this way the $2'' \times 2''$ aluminum angle is **enclosed** by the slot door on the front side of the dome when the door is **closed**. When the slot door passes the Slot Door Stop the back-side angle guide the slot door, centering it on the backside of the dome, preventing it from skidding off to the side. You will need two—8 ft sections of $2'' \times 2'' \times \frac{1}{4}''$ angle rolled R=5′–2″ Leg-out for the backside of the dome, and two 10 ft—6″ sections of $2'' \times 2'' \times \frac{1}{4}''$ angle Rolled R=5′–2″ Leg-out for the front side of the dome (slot door side). To complete the slot door guide assembly, you will also need a 3′ long $2'' \times 2'' \times \frac{1}{4}''$ linear slot door stop section which abuts the two 8 ft curved guides at the top of the front-side dome, and another 3′ long $2'' \times 2'' \times \frac{1}{4}''$ linear angle for the lower hatch area to install its hinge to the dome base ring. All of these components are bolted through the skin and underlying arches and braces with stainless steel flat-head $\frac{1}{4}'' \times 20$ bolts and nuts. Again, apply a good film of clear silicone sealant under and along each of the aluminum angle sections as you assemble them and bolt them down. You will need to do some fitting—requiring hack-sawing any edges that jut out too far, and filing to round off any sharp edges. Also, make sure that all the flat head bolts are flush with the surface of the outside surfaces or the slot door will rumble and bump on its passage over the dome.

This procedure will be required through the entire process of fitting the aluminum angle to the dome.

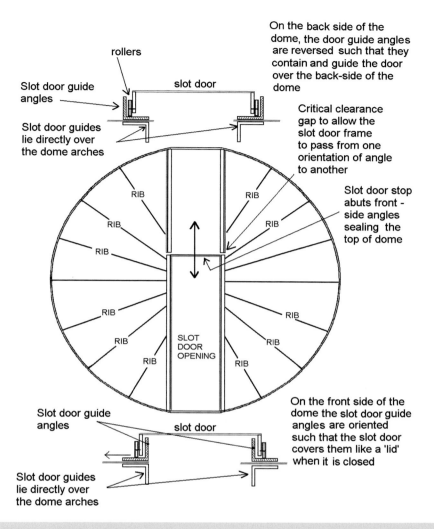

On the back side of the dome, the door guide angles are reversed such that they contain and guide the door over the back-side of the dome

rollers

slot door

Slot door guide angles

Slot door guides lie directly over the dome arches

Critical clearance gap to allow the slot door frame to pass from one orientation of angle to another

Slot door stop abuts front - side angles sealing the top of dome

RIB

RIB

RIB

RIB

RIB

RIB

RIB

RIB

RIB

RIB

RIB

RIB

SLOT DOOR OPENING

Slot door guide angles

slot door

Slot door guides lie directly over the dome arches

On the front side of the dome the slot door guide angles are oriented such that the slot door covers them like a 'lid' when it is closed

Fig. 15.34 Fitting the Slot Door guides

Construction and Operation of the Slot Door

Reaching this step in construction qualifies you as skilled because creating a dome in aluminum with only hand tools requires a degree of precision and effort beyond most undertakings in home craftsmanship. Working with wood allows some degree of error which can be readily corrected by laminating, gluing or patching, but aluminum joints and seams are unforgiving, requiring precise measurement and fitting.

The Slot door component is one such challenge, and perhaps the most critical in the whole assembly of the observatory. Your first function now that the dome is finished is to measure the exact distance (<u>circular</u> measure) from the top edge of the Slot Door Stop all the way down to

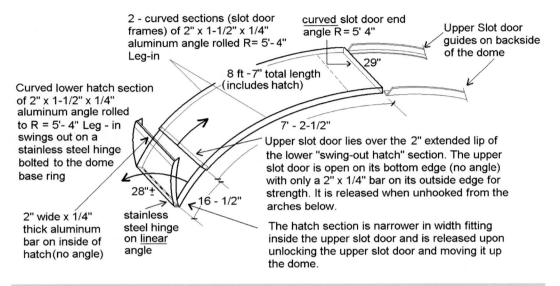

Note: The upper end of the movable section of slot door is terminated in curved angle, the lower end is linear with no angle - just an aluminum bar for strength

2 - curved sections (slot door frames) of 2" x 1-1/2" x 1/4" aluminum angle rolled R= 5'- 4" Leg-in

curved slot door end angle R = 5' 4"

Upper Slot door guides on backside of the dome

29"

8 ft - 7" total length (includes hatch)

Curved lower hatch section of 2" x 1-1/2" x 1/4" aluminum angle rolled to R = 5'- 4" Leg - in swings out on a stainless steel hinge bolted to the dome base ring

7' - 2-1/2"

Upper slot door lies over the 2" extended lip of the lower "swing-out hatch" section. The upper slot door is open on its bottom edge (no angle) with only a 2" x 1/4" bar on its outside edge for strength. It is released when unhooked from the arches below.

28"±

16 - 1/2"

2" wide x 1/4" thick aluminum bar on inside of hatch(no angle)

stainless steel hinge on linear angle

The hatch section is narrower in width fitting inside the upper slot door and is released upon unlocking the upper slot door and moving it up the dome.

Fig. 15.35 Slot door operation and construction

the Dome Base Ring surface. This will be the curved length of the Slot Door (both upper and lower hatch sections). Do this with a carpenter's tape from both sides of the slot door stop and use the longest dimension. From this, measure off the suggested length of the Lower Hatch adding an extra 2 in. to fit underneath the Upper section to act as a lock and weather seal. The curved length of the lower, swing-out hatch, should be 16-½″ plus the 2″ tongue that fits underneath the upper slot door section (18-½″ long in total, although only 16-½″ show on the slot door surface).

Depending on how far you went beyond the apex of the dome to accommodate ½ the width of your telescope objective, the balance or Upper Slot Door curved length will be the remainder between 16-½″ and the total distance you measured to the Slot Door Stop. In the model described herein that total distance will be 122″ making the Upper Door section 105-½ in. long in curved measure (i.e. 122 in. minus 16-½ in.= 105-½ in (see Fig. 15.35 for clarification).

If your measurement falls longer or shorter than that don't concern yourself because it will be a product of the accuracy to which the aluminum arches were rolled and the length you cut them to fit into the dome base ring. The width (outside dimension) of the Upper Slot door in our model here is 29″ to fit comfortably over the slot door frame which is 27″ I.D. on top of the 27″ wide arches underneath. The total width at 29″ O.D. includes a ¼″ thickness on both sides creating a tolerance of ½″ wide between the slot door guides and the slot door frame so that the door does not bind or jamb on the guides as it rolls past them.

The Lower Hatch must be narrower than the upper door such that when the Upper Slot door closes, it slides over the side angles of the lower hatch and its 2″ of "tongue".

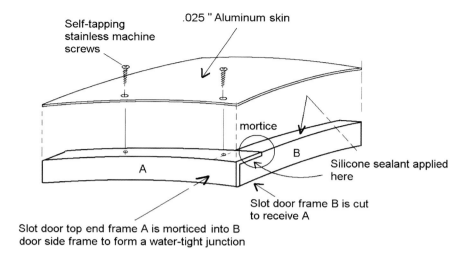

Self-tapping stainless machine screws

.025 " Aluminum skin

mortice

A

B

Silicone sealant applied here

Slot door frame B is cut to receive A

Slot door top end frame A is morticed into B door side frame to form a water-tight junction

Fig. 15.36 Fitting the corner angles of the slot door together

The Lower Hatch is carefully made to just fit over the leg of the slot door angle guides with it's I.D. 27-½″. The hatch O.D. will turn out to be 28″ with the width of it's frame angle at ¼″. That O.D. will fit within the 28-½″ I.D. of the Upper Slot Door when it slides down over it in closed position. Since the Lower Hatch I.D. is roughly equal to the Slot Door Frame angle's O.D. it will scrape it unless you add a 1/8″ allowance in measurement. Some minor alterations in construction will occur because any tiny errors tend to become cumulative at later stages, creating a lack of critical tolerance between overlapping structures, and structures that slide one along the other (remember that aluminum expands and shrinks with temperature).

Again, do not be alarmed if your measurements are slightly off these, the only important thing being that the parts fit the dome curvature and slot door guides, both in length, width, and with sufficient tolerance over the slot door guides such that the door slides over the dome and fits over the lower hatch comfortably.

The lower hatch is riveted or screwed with stainless self-tapping machine screws to a stainless steel piano hinge that swings from the dome base ring at its very bottom. In order to do this, the bottom of the lower hatch will have to be fitted with a linear aluminum angle. The top end of the hatch is left with only a reinforcing bar of aluminum along its underside (2″ wide × ¼″ thick), to fit under the upper door. The top end of the upper slot door is finished with a curved angle section which acts as a lid against the Slot Door Stop at the top of the dome. This curved angle also allows for an arc of exposed metal above the top edge of the linear door stop angle to locate a tie-down for the door. Aluminum is a difficult metal to use in a structure that requires sliding parts because they have a tendency to bind and grip if the tolerances are too small. In this respect do not try to make the slot door parts fit tightly to the guides for the sake of better weather sealing, because the system will not work effectively, and once made cannot be easily adjusted.

Fitting the corners of the slot door is tricky, in that one leg of each corner joint must be cut back to fit the leg of the other, as shown in Fig. 15.36.

The hatch must be suspended from its top edge by rope or chain—I found that light brass chain or stainless wire serves the purpose well, held at either end by stainless bolt eyes, threaded or bolted to both the slot door guides or the arches themselves. When closed, it is secured by the upper slot

Fig. 15.37 Photo of finished slot door and hatch on dome (omit skirt to be installed later)

Fig. 15.38 Photo of opened hatch suspended from light chains

door that overlaps its top edge. A double-ended boat snap clipped onto bolt eyes fastened on the slot door guide and the hatch secures it firmly at its top edge. Thus, the entire door is locked at its bottom (see Figs. 15.37 and 15.38 for lower hatch in closed and open positions).

On the backside of the dome you should install a spring-loaded arrestor to stop and hold the heavy slot door once it has rolled over the dome. Without this device, the heavy slot door will stretch your rope or cable, braking it suddenly at rope's end and loosening or damaging your pulley system.

Fig. 15.39 Photo of the spring-loaded arrestor on back side of dome

This part can be made easily from a small 8″ section of aluminum channel (wide enough to accomodate the slot door thickness). Alternatively, you could fabricate one from two sections of 1″×2″×2″ aluminum channel bolted together with a cut-off section of left-over ¼″×1-½″ rib as shown in Fig. 15.39. Two, long, ¼″×20 stainless bolts are run through heavy springs and bolted to the arrestor to prevent slamming of the dome. Place this device at the point where the slot door stops on the back of the dome.

The Slot Door Wheels

After trying several existing wheels from other systems (garage doors, various hard rubber casters etc.), I found only machined aluminum or brass solid wheels were suitable throughout all seasons. The wheels are best custom-made in a machine shop, although if you can procure a large diameter aluminum dowel, carefully cutting it into sections will suffice, sanding the faces smooth, and drilling a 3/8″ central hole to accommodate the bolt shaft. The 2-½″ diameter×0.460″ brass wheels must be located ½″ above the bottom edge of the slot door frame angle. This will provide the necessary clearance for the slot door, enabling it to roll easily along the exterior leg of the slot door guides.

Pay attention to the measurements in Figs. 15.40, 15.41, and 15.42 regarding clearances between the slot door frame and the slot door guides: do not reduce them.

The brass bushing that fits inside the machined wheel is necessary because it provides an excess length of 0.015″, which is the difference between the length of the brass bushing and the overall width of the wheel (0.475″−0.460″=0.015″). Looking closely at Fig. 15.40 you will see that there is also a 0.10 mm "hip" on the inside surface of the wheels which further prevents scuffing of the full diameter of the wheel against the slot door as it rotates. The "hip" on the wheel and the slightly-longer than wheel thickness brass bushing allows you to tighten the wheel bolt against the slot door frame, while allowing the slightly less thick wheel to rotate freely. This is the same mechanism that allows wheels on lawn mowers to rotate. These tolerances are all critical, tested on my prototype observatory, and must be adhered to.

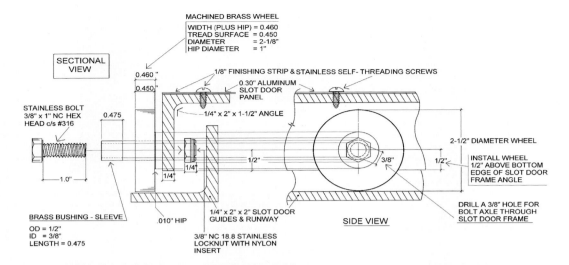

Fig. 15.40 Detail of slot door wheel assembly

A total of six wheels are required—a pair locate toward top and bottom of the upper slot door (but not precisely at the ends), and a pair locate mid-way between them. Placing the wheels at the very end of the upper slot door is not desirable producing too much pressure at the pull-down and pull-up points where the rope or cable is attached for opening and closure of the door. Offset the wheels from the slot door ends.

The clearance on the both the front side of the dome and the backside of the dome between the wheels and the door guides is critical, whether the door encloses the guides (on the front side of the dome) or the door is contained by the guides (as it rolls over the back side of the dome). If you position the slot door guide angles too close or too far apart, the door will bind. Either the axle bolt head outside the wheel will scuff the dome back-side guides, or the locknut securing the wheel inside the front door frame will scuff the front-side door guides jamming against them. Close adherence to the measurements given in the figure details will prevent this from happening.

The metal wheels will crush a thin ice layer and run through light snow accumulation, but anything over a few inches will hold any movement of door up over the backside of the dome. Because of their exterior location, the wheels have both a positive and negative aspect—easily accessible for oiling and tightening, but vulnerable to heavy snow and ice accumulation. However, I personally find that any depth of snow quickly slides off the dome after storms, usually in time for an observing session. A few light raps on the door with a straw broom usually dislocates any light snow.

The Slot Door Opening and Closing Mechanism

Most large commercial observatories have a motor—gear driven slot door system that opens and closes the slot door. In a small diameter observatory the need is not so great, and hand-operated systems will easily suffice, not needing the enormous mechanical advantage that larger heavier domes and doors require. Scaling the size down, smaller domes have a wide slot anyway in proportion to the diameter of the dome, and the open slot is much closer to the astronomer—easily accessible at almost arm's length.

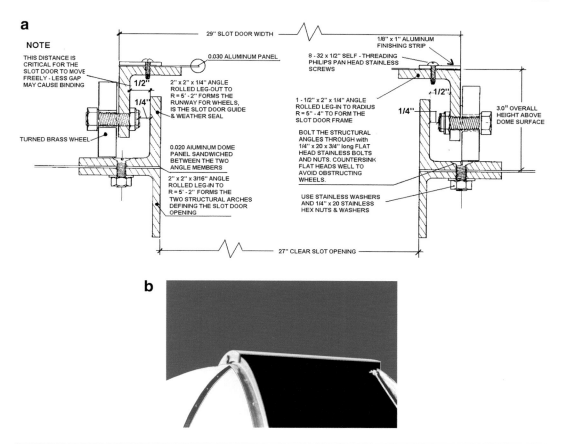

Fig. 15.41 (**a**) Detail of slot door shown on front side of dome in closed position. (**b**) Photo of slot door opening on front side of dome

The slot door width in this model is quite sufficient for almost an hour's viewing or astrophotography if you line up the object just along the left edge of the open slot door frame. It takes less than a minute to rotate the dome forward to gain another hour of astrophotography, and in this day of digital photography you will never need that long for any deep sky exposure. Hence, a simple rope and pulley system will fulfill years of service to move the slot door up and over the dome and back again. The system is designed to offer a 2× Mechanical advantage in either direction of travel—either to open or to close the door. The tension and friction in the pulley system will resist any tendency of the slot door to move on its own. You should be able to leave the door halfway open or at any position in its position of travel up and over the dome.

Using only your hands on the rope at the back of the dome, the door is easily moved through its entire travel avoiding the usual complex electrical worm drive gear system used in most larger observatories. This system was originally designed by Tatsuro Matsumoto in Japan for his fiberglass dome observatory, and is adapted here to suit this model (Fig. 15.43).

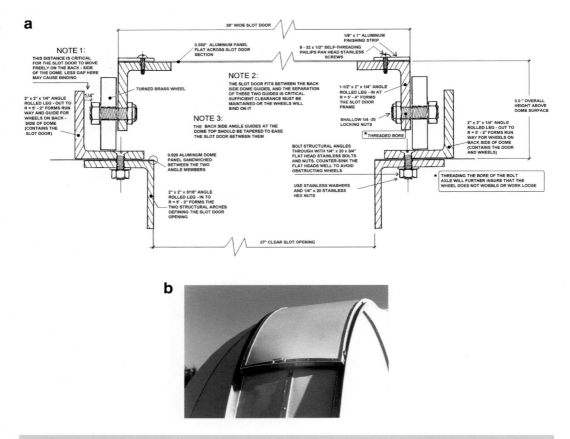

Fig. 15.42 (**a**) Detail of slot door shown on back side of dome (note reversed orientation of door guides). (**b**) Photo of slot door resting on back side of dome

Construction of the Elevated Observing Floor

Reasons for Elevating the Floor and Initial Planning

This observatory was designed with a full height entry door to allow easy access and to accommodate long focus refractors and reflectors. However, with the improvements in short focal length instruments, the extra long pier height may not be an advantage.

Also, the requirement for a ladder or rolling ladder to gain access to the telescope can be eliminated if the observing floor is constructed higher. The easiest way to support a new, higher floor, is by means of a curved structural angle rolled to the same radius as the radius of the inside wall, that will support a heavy joist floor and its occupants.

Spend some time locating a position for a stairwell that does not interfere with your observing position(s). Usually this will be in the most southern quadrant of your observatory. Also check out where a line of trees, or buildings, or light pollution threats might impair your viewing enough to sacrifice that sector of the observatory floor. The location will also be determined by the location of your entry door, and thus it is really important to figure this factor into the design and construction

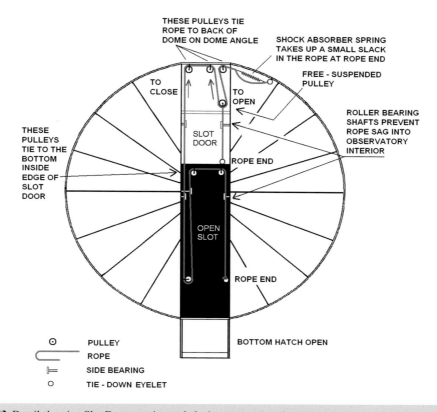

Fig. 15.43 Detail showing Slot Door opening and closing system (continuous rope)

before you begin to build this observatory. In my case, I was lucky, having located the door in the most southerly location of the observatory, not realizing that later on I would raise the floor. The southern location could easily hold a small stairwell being in a sector that I would normally not use for observing except for polar alignment, and observing the odd comet. Note that my pier was also offset from the central position which also allowed more room in the northern sector to observe southward. This is a detail that I would strongly suggest for short, low focal length, bulky reflectors, or Schmidt Cassegrain instruments, whose overall length is short enough that if you have to observe northward you can still fit in the area between the pier and the short distance to the south wall of the observatory. It is not necessarily convenient for long focus refractors that will squeeze you against the south wall when observing northward. Its not an easy decision to locate the pier off-centre should your choice of instruments change in the future, because the pier may or may not be in an undesirable position. Think this out carefully before you act.

I also suggest you paint the cement slab floor with a high quality epoxy resin—acrylic based coating for permanence and arrange all your electric outlets higher up above the proposed floor level. This will require more plastic conduit, wire, and receptacles (leave the existing ones in lower down as they will come in handy for vacuuming, temporary heating fans, or lights near the future stairs). Calculate where your finished floor will end up and place the electrical receptacles about a foot above the floor—on the wall and on the pier (you should by this stage already have a four outlet receptacle near the top of the pier).

Fig. 15.44 Photo of rolled steel angle protruding from joists on finished floor

There are other ways to support the joist floor, such as posts or joist hangers screwed to the individual studs all around the wall. Joist hangers unfortunately would have to vary in angle continuously around the inner wall making that route a carpenter's nightmare, but posts could be lag-bolted to the studs and an inner shelf constructed in arc-sections, from post to post, for supporting the joists. This method left less room for storage space below the floor with a much more complex wooden framework taking up extra room. To me, the rolled steel angle was so simple a solution that allowed for adjustment—the floor could be rotated (with a lot of effort) to slightly different positions before securing the stairs, to fix anything that had been overlooked, with a limited rotation around the offset pier.

Fabricating and Installing the Circular Steel Joist Support Angle

A 3″ × 3″ × 3/8″ thick steel angle rolled leg-in to the inside radius of the observatory will fit snugly against the inner wall to serve as a ledge for floor joists. If you built this observatory with the radial dimensions specified, that radius should be 4′–8-3/8″ exactly. Make sure you specify that it is rolled LEG-IN, and cut into four sections, 8 ft long for delivery and easier handling (3/8″ steel is heavy). You can specify that the steel fabricator drill ¼″ holes in the angle leg that is attached to the wall studs—16″ apart to match your stud separations before the angle is rolled. Once rolled in a bending mill, the heat and pressure will harden the steel and no ordinary metal drills will penetrate it, so it is imperative to have them pre-drilled. The best source for manufacturing such a steel angle is a farm silo manufacturer, because they use these circular steel angles at the base of their silo domes. DeMuth Steel Products in Canada and the USA supply this component for all their domes. To make attachment easier on the wall of the observatory you can ask them to cut oblong holes at 16″ on center in the Leg-up part of the angle to catch those studs that may be slightly less or more than 16″ apart (Fig. 15.44).

While the steel angle is being rolled, shop around for a new or used rolling ladder with a staircase that terminates at 45″ or so (i.e. the top platform). This will hold five steps, 7-½″ deep and 9 in. apart. Try to obtain one with hand railings, it will also have a full rail-surrounded platform at the top (they all do, for safety). It is important to select this now, as the exact stair height (from floor to platform on the ladder) will be your installation height for the steel angles. When you measure the staircase make sure the ladder is resting locked on the floor, not in its wheel-around condition (raised), as the little difference in height could produce an error in the height. Back at the observatory, mark this height around the inner wall with an erasable board marker—it will be the bottom edge of the 3″ × 3″ rolled steel angle. Now is a good time also to estimate just where the finished floor will end up using 2″ × 8″ joists resting on the rolled angle, plus the 2″ × 8″ flooring. That thickness should work out to be 9 in. Adding this to the 45 in. of stair height normally found on five step rolling ladders you will arrive at 54″ above the cement floor. If you want a perfect fit between the floor joists and the platform of the rolling ladder, include the 3/8″ thickness of the leg-in steel angle in your calculation making the overall height of finished floor 54-3/8″. Spray paint the rolling ladder and all the railings now (if necessary) before you begin to install them later, as painting them later will be awkward.

Upon delivery of your steel angle, paint the sections with two coats of metal primer and two coats of white rust preventative paint. Between applications, buy some light chain and heavy "S" hooks to hang the angle sections from your track to help you in bolting them to the wall. Hooking the "S" hooks (three per 8 ft angle) to the track edge, hang three sections of chain from each hook, placing an "S" hook at both ends and one in the middle of the angle section. The angle is heavy, so hook them temporarily at a convenient height, adjusting them back and forth until the angle bottom edge is where you have marked its location. When satisfied that it is in place drill the first ¼″ hole into a stud and screw in a 3″ × 3/8″ lag screw into the stud, tightly with a ratchet wrench. Go to the other end of the angle and do the same, then apply all the other lag screws. If you miss a stud do not be concerned, because 4 or 5 bolts per 8 ft length will easily hold the load.

You will likely have to cut the last section of steel angle to fit the door frame or perhaps to allow for a security siren, night light, or switch. Allow for this if you intend to install anything between the steel angle and the door frame(s) as it will be impossible to cut once installed against the wall.

Adding Joists, Bridging, and Flooring

Once all the circular angle sections are in place, mark on the angle where you want the joists to rest, and how many you want to install (see Fig. 15.45).

Note in the diagram that the joists run East–West with the stairway located south. This is the best orientation for most observatories unless your door is somewhere else, in which case you will have to add a trap door to cover the opening for observing purposes. It works out that exactly 6–2″ × 8″ joists are needed for a 10 ft diameter observatory, one of them a double joist at the stairway, and the others all single. The most northerly joist, a short one, lies almost entirely along the rolled steel angle.

At this point, before proceeding any further, cut the rolling ladder railings off the stair portion with a steel saw leaving them intact. Cut flush with the platform, and flush with the stair supports. Place the stairs in the observatory exactly where you want them located.

I'd advise strongly that they be placed where I show them to be in Fig. 15.45. Cut the first 2″ × 8″ double joist to fit along the stair's north edge in it's walk-in position near the door frame. Leave 1-½″ at this point between the joist and the pier. When satisfied that the stairs are in their correct location, add the bridging and short joist west of it and secure the stairs to its underside. The double joist will then be resting on the rolled steel angle, unattached, but firmly held secure by the stairs and bridge-work just installed.

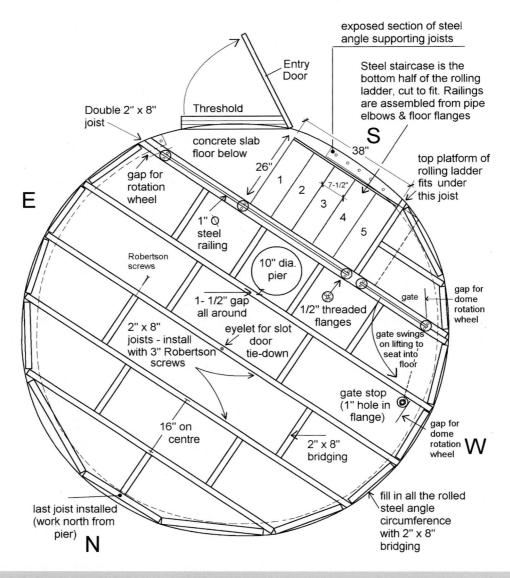

Fig. 15.45 Floor plan showing joists, bridging, stairway, and railing locations

Fill in all the joists on 16″ centers and bridging, securing with 3″ Robertson Flat head deck screws—use a cordless screw driver as hammering nails will be very hard in this enclosed environment. It should work out that all but the last joist will be 16″ apart.

Working northward, cutting each joist to fit the respective chords of the observatory wall circle, install 2″ × 8″ bridging as you go. The second joist should again be 1-½″ from the pier as should the two bridges that surround the pier. Pre-marking the joist positions on the rolled steel angle with a magic marker will make the job easier. Lightly nail bridging all around the circumference between the joists (toe-nail into the joists) to support the floor at its edges. When complete, begin to screw down the 2″ × 8″ board floor starting at the pier working east and west across the joists. Use 3″ Robertson flat head screws and counter-sink the heads into the boards, leaving a 1/8″ gap between

Fig. 15.46 Photo showing the staircase in place under the finished floor

floor boards for expansion on hot days. You will find it necessary to make a template for each end of the board to fit the curvature of the wall because the arc of the wall varies as you proceed around the circular angle. Cut cardboard with scissors to fit and transfer it to the board ends. Make sure you measure, and mark, the longest distance from wall to wall on each board and then apply the template to meet the end of that mark. Cut the curved ends with a portable hand-held jig saw. When finished installing the floor, paint it with two coats of high quality spar varnish (it will sooner or later get wet from a storm) (Fig. 15.46).

Stairs and Safety Railings

The railings that you cut free from the staircase of the rolling ladder can be installed at this time on the finished floor. Use ½″ plumbing threaded pipe (internal diameter) to join sections of railing together. Insert the pipe up into the railing and drill through the railing and pipe, bolting them together with 1/8″ S/S bolts. The railings are easily attached to the floor using the same process and secured with ½″ threaded pipe floor flanges. These are quite rigid when screwed to the floor, the ½″ pipe pre-threaded into them. The former rolling ladder railings should be enough to fabricate a set of handrails—one long hand rail stretching across the edge of the floor for safety near the open stairwell, and a short section to create a gate at the top of stairs. The long hand-rail will have to curve downward well before it reaches the vicinity of the telescope in order to accommodate the swing of the equatorial mount, and is permanent in position. It will have to taper downward as it approaches the telescope mount such that it allows the telescope to swing in its arc of travel past it. The other end must leave sufficient clearance for the dome rotation mechanism as it swings around with the dome.

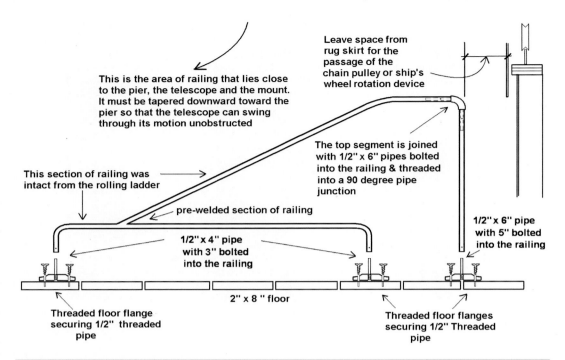

This is the area of railing that lies close to the pier, the telescope and the mount. It must be tapered downward toward the pier so that the telescope can swing through its motion unobstructed

Leave space from rug skirt for the passage of the chain pulley or ship's wheel rotation device

The top segment is joined with 1/2" x 6" pipes bolted into the railing & threaded into a 90 degree pipe junction

This section of railing was intact from the rolling ladder

pre-welded section of railing

1/2" x 6" pipe with 5" bolted into the railing

1/2" x 4" pipe with 3" bolted into the railing

2" x 8 " floor

Threaded floor flange securing 1/2" threaded pipe

Threaded floor flanges securing 1/2" Threaded pipe

Fig. 15.47 Fabricating the long railing at the edge of the elevated floor

Follow the diagram to cut lengths of ½″ pipe and the position of the floor flanges. You must obviously turn the floor flanges onto the threaded ends of the ½″ pipe before bolting the other ends into the hand-railing. The whole length of hand-rail excepting a short section of the bottom "kick-rail" will come intact off the old rolling ladder frame. Create the lower "kick-rail" by joining a short 90° section of railing to the longer section with a bolt-in ½″ joiner pipe (Fig. 15.47).

The short section of hand rail forms a swing-out gate at the top of the stairs, closed when the observatory is in operation (preventing people from falling down the stairs). The swing-out hand-rail gate comes complete from the top-end section of the former rolling ladder (it forms the end of the "cage" surrounding the former stair platform at the top). Using this intact section, hold it over the floor along the same axis as the long hand-railing, at the top inside edge of the staircase, and mark the tube centers of both ends on the floor surface. Drill a 1″ diameter hole through the floor on the marked center near the telescope, and locate a ½″ floor flange over it with its thread drilled out to a diameter of 1 in. Locate another floor flange over the opposite marked center nearest the wall <u>with its threading intact</u>.

Test the positions of hole and floor flange by inserting a ½″ pipe up into the tubing (tape it in position temporarily) before drilling. This will locate your swing-out gate. Make sure that you again leave sufficient space at the wall to allow the dome rotation device to swing through it. You can add another drilled hole at the far swing position of the open gate if you wish, as a secure rest position for it, until you wish to close it. Again, make sure you leave sufficient clearance for the dome rotation mechanism at this point. Thread a 6″ length of ½″ pipe into a floor flange nearest the wall and loosely insert the pivoting side of the gate into it—this is the bearing point. The other end of the gate tube will take a permanent ½″ pipe, 4 in. in length with 2″ bolted into the tubing—this is the pipe that will lift out of the floor flange nearest the 'scope and swing out of the way to rest near the wall (in its far swing position). To open the gate, it is lifted out of position and swung out of way 90° into its open position,

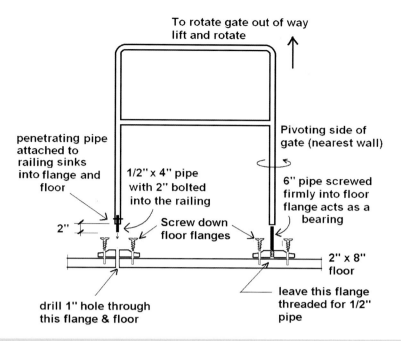

To rotate gate out of way
lift and rotate

penetrating pipe
attached to
railing sinks
into flange and
floor

1/2" x 4" pipe
with 2" bolted
into the railing

Pivoting side of
gate (nearest wall)

6" pipe screwed
firmly into floor
flange acts as a
bearing

2"

Screw down
floor flanges

2" x 8"
floor

drill 1" hole through
this flange & floor

leave this flange
threaded for 1/2"
pipe

Fig. 15.48 Fabricating the swing-out stair gate

its fixed pipe penetrating the floor flange and hole in the floor. You will need to calculate what length to cut the gate tubes off since the lengths are NOT equal for the gate to work properly. Work this out (Figs. 15.48 and 15.49).

Cover the floor in a good quality rugged indoor/outdoor carpet, and add a chair mat for rolling back and forth to the computer.

Lifting the Dome onto the Track, Securing the Dome, and Adding the Skirt

Lifting and Securing the Dome

You can lift the dome (still resting on blocks on the ground I hope) with the aid of eight friends—at least with a lower 4 ft high wall observatory, but you better forget about this method with a high-wall model. To guarantee safety and the assurance of seating the casters on the track, hire a light crane. If you have followed my suggestion in the first section of the book, you will have provided access for heavy equipment to do this, or make way for such. A standard truck crane is well worth the hourly cost to lift your precious dome up onto the track. Some people can achieve this with a back-hoe if the operator is careful. Follow the hook-up point locations shown in Fig. 15.50.

Be prepared to rig up your own lines well before the crane arrives, as crane time is expensive—you want to have everything ready. Using strong nylon rope (sailing sheet rope is fine), rig up an inverted parachute-like basket that terminates outside the top of the dome in a strong ring or hook. Following the details in Fig. 15.50, you should end up with six heavy-duty bolt-eyes equidistant around the dome plywood ring. Bolt the eyes right through the plywood layers being careful to position them offset from the six casters.

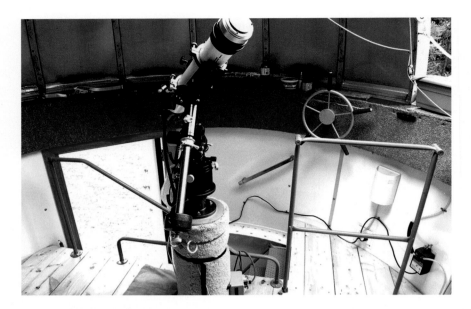

Fig. 15.49 Photo showing the two finished railings in place

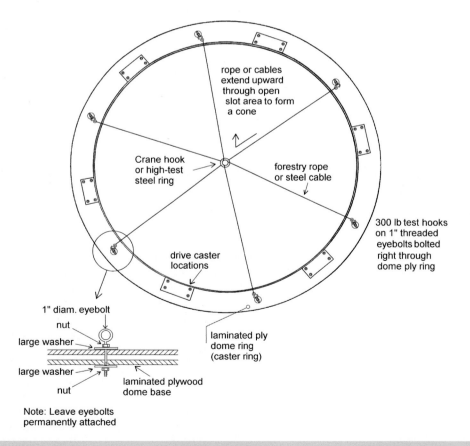

Fig. 15.50 Pick-up points for crane-lifting the finished dome

Use large diameter, thick washers underside to spread the lift force over a larger area of plywood. When you are satisfied all the leads terminate equally to the ring or hook above the dome, tie a separate length of rope to the dome around the plywood ring at the slot door to guide the dome with as it is lifted (refer back to Fig. 13.3 to see Jack Newton guiding the dome with a rope line). This is necessary since the crane operator most likely won't be able to see the subtle seating of casters onto the track from his cab. When finally in place, rotate the dome by hand to make sure it rotates freely. If it does not do so, retain the crane operator, and with a step ladder undo the heavy screw(s) that tie down the track in any area that seems to be binding the casters. Estimate how much bending is required, and push or pull that area in or out (use a large wood clamp to do so), and re-set the screw(s) . You will find the track will give easily over a small distance. This should not actually happen as you tested the rotation of the bare dome ring complete with casters much earlier in the process. Then, let the crane operator go as soon as possible.

Once the lifting operation is complete, install a heavy 3/8″ S/S eyebolt towards the top edge of the inner curved end of the upper section of slot door—in a central position. You will need a step ladder to do this, place it mid-point in the observatory on your new floor. Climb up and Drill a 3/8″ hole through the curved aluminum end of the slot door for the eye-bolt and attach a stainless nut on the other side. This will be your dome lock-down point. You might at this stage appreciate why the upper end of the slot door was curved. The door stop at the top outside edge of the dome is linear, and when the slot door is closed, its curved end comes up against the linear leg of this door stop—leaving just sufficient gap for placing an eye-bolt in the segment above it. I've shown the eye-bolt face on so you can see it, but it projects toward you from its location (see Fig. 15.51 for clarification).

To complete the tie-down apparatus, using a tape measure, hook onto the eye-bolt and measure the distance to the 2″×8″ floor surface. Drill through the floor planks where the tape reel ends and insert a heavy-duty eyelet at least 3/8″ thick. Insert a large washer on the underside and topside to spread the force on it. Fabricate a solid tie-down tube from a section of aluminum tubing with eye-bolts fastened on both ends (3/4″ diameter tube should suffice). It should be long enough to leave maximum 24″ clearance from the floor.

You will have to make some plugs to fit the tubing in order to attach the threaded eye-bolts. Insert solid ends (already drilled through for the eye-bolts) into the tube, and pin or screw the ends firmly into the tube. Next, fasten a short (24″ max.) section of stainless steel braided wire to a snap in the end of the tube's bottom eye-bolt and loop it through another heavy-duty sailing snap at the lower end of the stainless wire. Cable clamp the wire ends together after adjusting the length to just be able to clip the tie-down to the floor eye-bolt without difficulty—it should however, be tight on clipping it.

The last part of this fabrication is tricky—the stainless wire rig has to be slightly flexible enough to snap it down over the eye-bolt on the floor. Attach a standard bungee cord between the eyebolt on the tubing and the lower snap that's to engage with the eye-bolt on the floor. If you have figured out the length of the stainless wire correctly relative to the bungee cord, you will be able to just snap the tie down apparatus to the floor by extending the bungee cord only a fraction of an inch. This will take some adjustments to get it perfect. The eye-bolt on the top end of the tube will have to be cut to form a hook to easily engage the eye-bolt at the inner top of the slot door. To secure the dome and door (the door pulls down on the dome by this process) reach up and hook the upper eye-bolt (now cut like a hook) into the eye-bolt on the slot door frame end, then fasten the lower end of the tube's stainless wire snap onto the floor eye-bolt (Fig. 15.52).

This prevents any storm damage to the dome and slot door . In high winds, the dome creates an airfoil, allowing wind to both go through the track gap and also over the dome creating lift (due to its length of travel). Many observatories lose their dome in high winds because the owners did not storm-proof them (look back at Fig. 15.15).

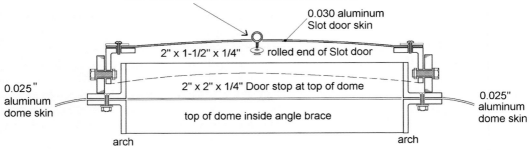

Eye - bolt penetrates rolled angle at top end of Slot
door to tie down the dome & Slot door to the floor

0.030 aluminum
Slot door skin

2" x 1-1/2" x 1/4"   rolled end of Slot door

2" x 2" x 1/4" Door stop at top of dome

0.025"
aluminum
dome skin

top of dome inside angle brace

0.025"
aluminum
dome skin

arch   arch

Notice that the slot door stop at the top of the dome
is a <u>linear angle</u>, the slot door frame end is a <u>curved</u>
angle

The difference in shape between these two members
creates a gap at the top of the dome for a tie - down
eye bolt fixture. This connection has an advantage in
that the slot door must be shut to gain access to it,
and when clipped to the floor, pulls down on both
dome and slot door anchoring them to the observatory
floor. It is the only point capable of providing protection
for both.

Fig. 15.51 Detail of top end of slot door from inside showing lock-down eye-bolt access above the slot door stop angle

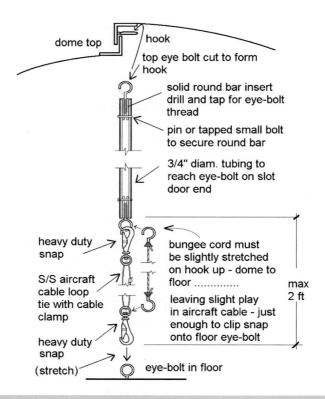

dome top   hook

top eye bolt cut to form
hook

solid round bar insert
drill and tap for eye-bolt
thread

pin or tapped small bolt
to secure round bar

3/4" diam. tubing to
reach eye-bolt on slot
door end

heavy duty
snap

bungee cord must
be slightly stretched
on hook up - dome to
floor ....

S/S aircraft
cable loop
tie with cable
clamp

leaving slight play
in aircraft cable - just
enough to clip snap
onto floor eye-bolt

max
2 ft

heavy duty
snap

(stretch)   eye-bolt in floor

Fig. 15.52 Detail of dome tie-down tubing and fixtures

I added extra protection by installing two seat-belt type ratchet tie-downs on opposite sides of the dome. These are clamped to two of the eye-bolts in the dome plywood base and eye-bolts screwed into a wall stud above the new floor (refer back to Fig. 15.49 — the yellow ratchet tie-downs are visible lying on the dome plywood ring).

To stop any rotation by wind I also installed a large turn-buckle bolted through the base of the track, which when raised up, aligns with an eye-bolt screwed into the bottom of the plywood base ring. Bolted together with a short, heavy bolt and nut, it prevents any movement of the dome.

In its new position, the telescope may need a pier extension to make a comfortable sitting position. You will be pleasantly surprised to find the dome now within reach of your out-stretched arm. The dome rotation system, now at waist level, is completely accessible.

The observing floor will be less damp now, high above the cold cement floor, and the dome exterior available from inside requiring no more than a step-ladder on the floor. The slot door opening and closing mechanism is also at waist level, and the swing-out hatch pleasantly within reach. An extra observing bonus is that the increased height of the observing floor is slightly warmer under the dome's proximity.

Adding the Skirt

To apply the skirt around the base of the dome, you must first remove all the self-tapping S/S screws at the bottom of the dome skin where it terminates over the dome plywood ring. This does not include the rib bottoms, and the arch bases that were fastened to the dome base angle with flat-head, recessed bolts — they are just covered over.

You need to be able to shove the skirt panels up under the skin, over the arches and rib ends, and then re-screw the skirt in place. Measure the gap between the top of the door and the termination of the dome skin on the Dome base angle — that distance minus a tolerance of about an inch will provide a deep skirt that misses the door when opened.

My home model ends up to be 10 in. of skirt exposed with 2″ under the dome skin Cut four sections of skirt about 8 ft long × 1 ft wide to cover the 32 ft of circumference from 0.030 in. thick aluminum sheet. This gauge of aluminum is much stiffer than the dome skin above it, since it has a free-standing bottom edge.

Begin by shoving the first panel under the skin for its whole 8 ft length — make sure it is not buckled and insert a S/S screw in either end. You may have to re-drill through the former screw holes to pierce the skirt. When satisfied the first skirt panel is fixed in place, re-screw all the dome skin back in place over it. Overlap the panel leading edges about an inch and proceed in the same manner screwing all the panels down. Measure the depth of skirt all around to make sure it is even as you install it — you should have about 2″ of skirt concealed under the dome skin and ribs, with 10″ hanging below. All the screws should turn into the plywood laminated ring beneath the dome and arches. The over-lapped edges of skirt between each 8 ft section are bolted together with light 10–24 S/S machine bolts. The finished skirt should have a 2″ gap between the circular wall and the inside edge of the skirt — leaving enough room to get your hand up to reach the track. You may find later on that you want to install a foam or brush weather-strip in this gap to reduce draughts (refer to Fig. 15.32 for details on installing the skirt).

Add a final finishing strip around the perimeter of the skirt, if you wish, for cosmetic purposes. A strip of 1/8″ × 1″ light aluminum bar creates a nice edge to an otherwise rough-cut skin edge, but it requires removing many of the existing S/S screws again around the dome bottom, since you cannot fit this over round-headed screws you just installed. It serves two purposes actually, adding a "finished look" to the dome and it compresses the layers of skin and skirt underneath creating a better weather-seal. Adding a generous strip of silicone caulking on the upper edge of this strip prevents any water seeping under it.

Weatherproofing the Structure

At his point you should have applied silicone caulking to all the rib seams, arch seams, slot door guides, slot door angle and edges of angles forming the hatch. The entry door itself will eventually become a problem area because there is little overhang above it, allowing rain to seep into the interior or snow to pile up against it. The threshold is a constant problem being too close to the ground where ants seem to want a home. Many species will eat their way through the door jambs becoming a nuisance. Because the face of the door faces south, it intercepts most of the summer storms, and driving rain, and no weather-strip I've tried so far seals it effectively. I rolled a round cement well top into an excavated soil cavity, to lie flush on the grass as a small stoop in front of the door.

I then cut a plywood sheet to fit the width of the doorframe to rest on the stoop, angled against the door. This arrangement prevented any rain from driving in the door at the threshold and it also discouraged ant nests next to the door. I chose a cement well top because it was round and mirrored the shape of the dome.

The dome top can become another leakage point. In heavy wind storms, the marriage point between the dome stop angle and the slot door end may admit a little water. It is almost impossible to seal this area better, as any weatherproof material (foam, rubber etc) will scrape against the moving door as it door rolls over the back of the dome. My solution was to make a "curtain" that hung from curved stainless tubing alongside both inner arches which caught any water admitted from either dome top or the gap between the slot door and the dome. This ran any water collected downward into a linear plastic container alongside the inner hatch. A light sash rope rigged on pulleys allowed it to be pulled up to the top of the dome for observing, easily released on closing the dome up (Fig. 15.53).

The inside caster gap should be covered in an attractive rug material that should match the rug you install on the observing floor (for cosmetic reasons). Pre-cut panels of rug should be cut to fit the width of the caster gap (measured from top of the dome plywood ring to the bottom edge of the wall's upper plywood ring). These are screwed to the upper dome plywood ring with S/S round-headed screws and finishing washers on 1 ft intervals.

The leading edges of the rug (where each panel meets another), must be sewed together with heavy nylon line. When you come to the chain pulley or drive, you will have to cut the rug to fit around it. (Take the pulley off, cut the rug to fit over the shaft, and then replace the pulley).

The edge of the vinyl wall covering, hanging over the base, should be weatherproofed where it hangs over the cement slab. Ants, centipedes, stink-bugs, and lady-bugs will crawl up the vinyl seams, particularly in the battens, and end up eating the laminated plywood base ring where you cannot see it. Spread a heavy silicone seam all along the bottom edge of vinyl all around the slab. If there is any exposed cement below the vinyl siding, apply two coats of cement sealer to it. Also make sure the seams around the outside electrical box are well-sealed with silicone.

Fig. 15.53 Photo of the slot door curtain for waterproofing

Final Luxuries and Interior Fixtures

If you found that the slot door opening and closing mechanism shown in Fig. 15.43 does not suit your layout you can adopt a simpler system shown in Fig. 15.54.

This system runs on two rope haul systems rigged on either side of the interior Arches. The "haul up" pulley operates on a fiddle/becket block pulley situated about ¾ way up the inner leg of an arch. This pulley flips over as the rope passes it, and the slot door travels further to the top of the dome. When it travels downward this rope again causes the pulley to flip over acting as a bearing to lower the slot door. The other rope which is only used for "haul down", passes over a free-standing pulley on a shaft, that has a "keeper" on it which retains the rope, preventing it from dislocating every time you lower the slot door. The two ropes can be tied together at the bottom with sufficient slack to prevent them from dangling down into the observatory. It works very well—perhaps with less effort than the Tatsuro design.

The chain drive dome rotation wheel can be replaced by a small ship's wheel if you can find one, but they are difficult to find. With the elevated floor, you no longer need the chain drive and grooved pulley to turn the dome. A simple ship's wheel is very accessible at waist height and turns the dome with ease. It also adds 'panache' to the observatory. The wheel center must be drilled out to a diameter of 1 in. (same as the former chain drive pulley axle). Drill a small set-screw hole at right angles to the axle and thread it for a 3/8″ allen screw. Before you insert the allen screw, drill a ¼″ hole right into the axle to indent it so that the allen screw will not slip on the shaft and will bite' into it (Fig. 15.55).

The pier above the floor can be wrapped in the same carpet that you used on the floor, This is a good way to retain a degree of comfort on those cold nights when your legs rest against the pier, and its much more attractive than an unfinished aluminum pier.

Fig. 15.54 Photo of simple Slot door top pulley system (alternative to Fig. 15.43)

Fig. 15.55 Dome rotation wheel made from a ship's wheel

Fig. 15.56 Photo of pier wrapped with rug

Cut the carpet to wrap fully around the pier, so that its ends just meet, and sew them together with nylon fishing line. Cut out any areas to fit the carpet around electrical boxes or piping (Fig. 15.56).

A small wall-mounted lamp with a shield to prevent glare is a real asset if supplied with a 25 W red bulb. This will allow you to see your way around at night but won't spoil your night vision. Again, make sure the chain wheel or ship's wheel doesn't hit it on rotation.

A small table for your computer, star maps and notes, can be rigged on the curved wall at your waist height. If you use a section of $3'' \times 3'' \times 3/8''$ rolled angle left over from the floor construction, it has a radius of curvature to fit the inner wall. A plywood top, curved to match the wall, will not interfere with your activities and is easily supported by the rolled angle on the wall, and two large right-angle metal supports. Finish this off with a piece of counter-top laminate, in Starry night décor, contact-cemented to the plywood.

Last but not least is an alarm system. I prefer an infra-red alarm system which activates on breaking the infra-red light beam. Mounted on an inner door frame with its reflector placed on the adjacent inner door frame, it catches anything that opens the door and passes through it. A medium size horn is placed just below the alarm system, tucked more inside so it cannot be seen easily on entering. It is deafening on tripping, with the blast rebounding inside the dome and walls—guaranteed to remove any intruder (Figs. 15.57 and 15.58).

You may need a ventilating fan system to cool off the dome and area above the elevated floor. This should go below the elevated floor because if it is reversed to blow outward (towards the exterior of the observatory) it will pull air DOWN through the open slot door reducing convection currents above your telescope. At night it will draw cooler air in through the door frame to cool off a warm observatory. Refer back to Chapter 11 (Figs. 11.4 and 11.5) for details.

Fig. 15.57 Photo of infra-red alarm security system

Fig. 15.58 Photo locating infra-red alarm on inner door frame (*circled*)

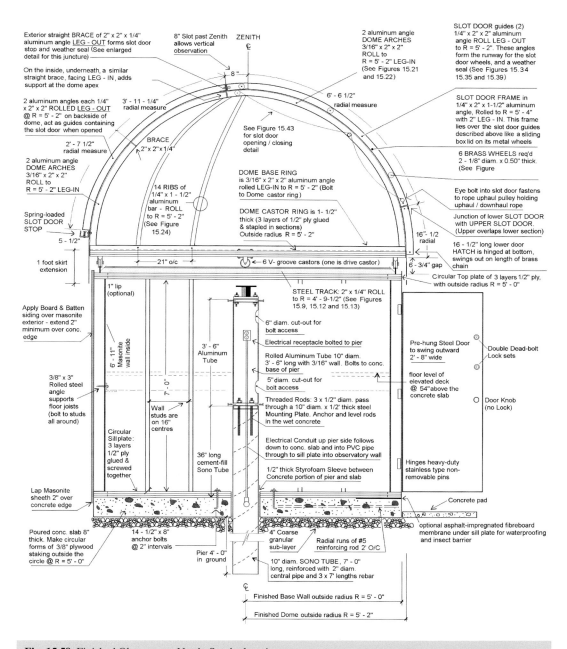

Fig. 15.59 Finished Observatory North–South elevation

The completed dome observatory is illustrated in the following two drawings (Figs. 15.59 and 15.60). It is worth referring to these two as your project progresses to see where things fit and as a check on material specifications and sizes. Note that the elevated observing floor is shown in hatched line as an add-on, if desired. I found it was necessary to access the dome, and because I had replaced my long focus refractor with a short, low f/ratio refractor. If you have a long focus instrument, think twice about installing the elevated observing floor. These two drawings summarize and end the process of building a dome observatory.

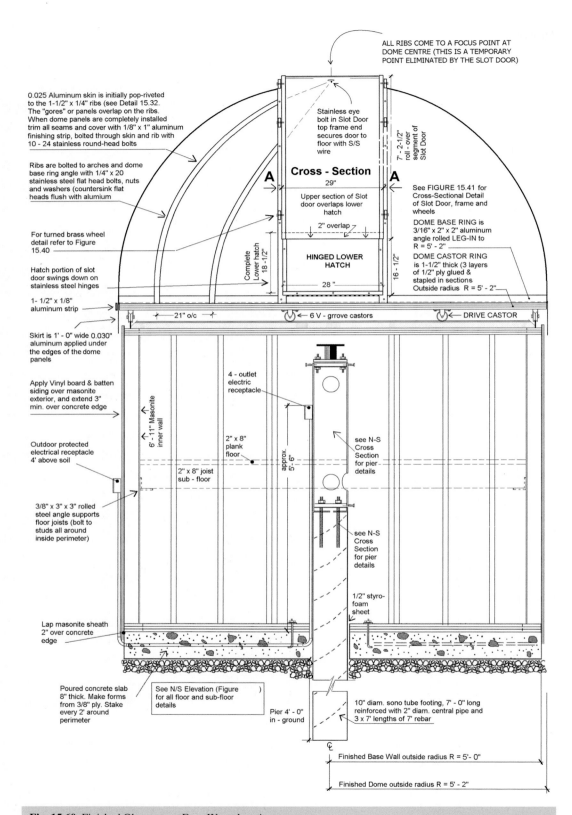

ALL RIBS COME TO A FOCUS POINT AT DOME CENTRE (THIS IS A TEMPORARY POINT ELIMINATED BY THE SLOT DOOR)

0.025 Aluminum skin is initially pop-riveted to the 1-1/2" x 1/4" ribs (see Detail 15.32. The "gores" or panels overlap on the ribs. When dome panels are completely installed trim all seams and cover with 1/8" x 1" aluminum finishing strip, bolted through skin and rib with 10 - 24 stainless round-head bolts

Stainless eye bolt in Slot Door top frame end secures door to floor with S/S wire

7 - 2-1/2" roll - over segment of Slot Door

Ribs are bolted to arches and dome base ring angle with 1/4" x 20 stainless steel flat head bolts, nuts and washers (countersink flat heads flush with aluminum

Cross - Section

29"

A   A

Upper section of Slot door overlaps lower hatch

See FIGURE 15.41 for Cross-Sectional Detail of Slot Door, frame and wheels

For turned brass wheel detail refer to Figure 15.40

2" overlap

16 - 1/2"

DOME BASE RING is 3/16" x 2" x 2" aluminum angle rolled LEG-IN to R = 5' - 2"

Hatch portion of slot door swings down on stainless steel hinges

Complete Lower hatch 18 - 1/2"

HINGED LOWER HATCH

28 "

DOME CASTOR RING is 1-1/2" thick (3 layers of 1/2" ply glued & stapled in sections Outside radius R = 5' - 2"

1- 1/2" x 1/8" aluminum strip

21" o/c   6 V - grroove castors   DRIVE CASTOR

Skirt is 1' - 0" wide 0.030" aluminum applied under the edges of the dome panels

Apply Vinyl board & batten siding over masonite exterior, and extend 3" min. over concrete edge

6' - 11" Masonite inner wall

4 - outlet electric receptacle

2" x 8" plank floor

see N-S Cross Section for pier details

Outdoor protected electrical receptacle 4' above soil

2" x 8" joist sub - floor

approx. 5'-6"

see N-S Cross Section for pier details

3/8" x 3" x 3" rolled steel angle supports floor joists (bolt to studs all around inside perimeter)

1/2" styro-foam sheet

Lap masonite sheath 2" over concrete edge

Poured concrete slab 8" thick. Make forms from 3/8" ply. Stake every 2' around perimeter

See N/S Elevation (Figure   ) for all floor and sub-floor details

Pier 4' - 0" in - ground

10" diam. sono tube footing, 7' - 0" long reinforced with 2" diam. central pipe and 3 x 7' lengths of 7' rebar

Finished Base Wall outside radius R = 5'- 0"

Finished Dome outside radius R = 5' - 2"

Fig. 15.60 Finished Observatory East–West elevation

Fig. 15.61 Photo of Finished Observatory looking north

The observatory location against a beautiful pine forest, enhances its structure and contributes to its effectiveness. The pine trees (conifers) produce less air turbulence above their crowns on hot days than deciduous trees, due to less transpiration of water vapour from their narrow needles. This creates a better atmosphere above for solar observing, should that be your passion in astronomy. Open fields also add to the turbulence factor with a higher reflectivity (albedo). The forest creates a more stable air column for observing and photography (Fig. 15.61).

For comparison, the very affordable commercial models might be your choice of observatory, particularly if you lack the necessary construction skills to build the model I have outlined here. The SkyShed POD, Home Dome, or Pulsar models manufacture very fine domes, and most suppliers will ship you a dome unit only. You can arrange for a carpenter to make your base with the plans in this book. Make sure to show the supplier the intended portion that you want to make, along with measurements before you begin, because diameter is critical (Fig. 15.62).

Maintenance

A schedule should be established for the following:

(a) lubricating the V-groove roller bearings—once per year
(b) spraying the aluminum dome with aluminum or silver spray—every 4 years
(c) spraying against insect invasion (ants) around the cement pad perimeter—once a week
(d) painting the inside face of the steel track—every 4 years
(e) painting the exposed surfaces of the plywood dome rings—every 4 years

Inspect the cement floor for cracks and erosion (the freeze-thaw cycle pulverizes cement surfaces) and parge with resin-cement. If wasps or hornets become a problem, locate several small canisters of moth balls around the dome base ring beside the track. (They are not good for humans so remove

Fig. 15.62 Photo of SkyShed POD Observatory and it's proud owner [Richard Kelsch, Ontario]

them when entering.) It is vital to keep ahead of their establishing nests, preventing a sting. Keep a constant eye out for leaks after heavy rain storms and re-caulk the seams where drops come from (you will have to weather a storm inside to find the exact spots). Be aware of tree growth around the observatory—even high boughs—squirrels can leap off boughs, land on your dome and get inside the slot door seams, building a nest and chewing computer and drive wires! Transplant any young trees you have to move. Move them before they get big and you cannot move them. Prune or cut down trees that specifically block or appear to start blocking your views. Periodic maintenance will keep your observatory in top shape and it will last your lifetime of observing.

Finishing the observatory was achieved without ceremony, hand-shaking or speeches. It was a silent celebration between myself and the heavens. I did, however, feel like I had created a machine of sorts that was going to allow access to new discoveries—and I couldn't wait to use it. It has been like that ever since. An observatory is a doorway between you and the universe. Your telescope is the access to it. Good luck.

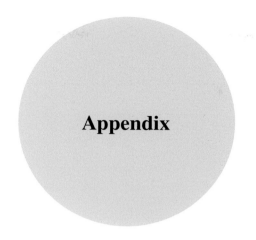

Appendix

Nails

1″	Ardox	2d
1-¼″	Ardox	3d
1-½″	Ardox	4d
1-¾″	Ardox	5d
2″	Ardox	6d
2-¼″	Ardox	7d
2-½″	Ardox	8d
2-¾″	Ardox	9d
3″	Ardox	10d
3-¼″	Ardox	12d
3-½″	Ardox	16d
4″	Ardox	20d
4-½″	Ardox	30d
5″	Ardox	40d
5-½″	Ardox	50d
6″	Ardox	60d

© Springer-Verlag New York 2016
J.S. Hicks, *Building a Roll-Off Roof or Dome Observatory*, The Patrick Moore
Practical Astronomy Series, DOI 10.1007/978-1-4939-3011-1

Metric Equivalents: Wood Construction

Board Sizes

Imperial	Metric
1″ × 2″	20 × 38 mm
1″ × 3″	20 × 63 mm
1″ × 4″	20 × 90 mm
1″ × 6″	20 × 150 mm
1″ × 8″	20 × 180 mm
1″ × 10″	20 × 240 mm
2″ × 2″	38 × 38 mm
2″ × 4″	38 × 90 mm
2″ × 6″	38 × 150 mm
2″ × 8″	38 × 180 mm
2″ × 10″	38 × 240 mm
2″ × 12″	38 × 275 mm
4″ × 4″	90 × 90 mm
4″ × 6″	90 × 150 mm
4″ × 8″	90 × 180 mm
6″ × 6″	150 × 150 mm

Metric Equivalents

PLYWOOD: Same Spec. for Waferboard/Particle Board/Masonite

Imperial	Metric
1/8″	3.00 mm
¼″	6.00 mm
3/8″	9.50 mm
½″	12.00 mm
5/8″	16.00 mm
¾″	19.00 mm
1″	25.00 mm

Steel Track Angle: Roll Off Roof

1-½″ × 1-½″	38 × 38 mm (knife edge track as suggested)
1″ × 2″	20 × 50 mm (flat track alternative)

Steel Track: Dome

2″ × ¼″	25 mm × 6 mm
½′ × ½′ bar	12 mm × 12 mm

Aluminum Structural Angle: Dome

2″ × 2″ × 3/16″ aluminum angle	50 × 50 × 8 mm
2″ × 2″ × ¼″ aluminum angle	50 × 50 × 6 mm
2′ × 1-½″ × ¼″ aluminum angle	50 × 38 × 6 mm
1-½″ × ¼″ aluminum bar	38 × 6 mm

Steel Angle: Observing Floor Support—Dome

3′ × 3′ × 3/8″ steel angle	76 mm × 76 mm × 9.5 mm

Aluminum Panel: Dome

8′ × 10′ × 0.025″ aluminum sheet	2438 mm × 3050 mm × 0.635 mm
8′ × 10′ × 0.030 aluminum sheet	2438 mm × 3050 mm × 0.762 mm

To convert	Multiply by
Sq. m to sq. yd	1.196
Cu. m to cu. yd	1.308
Sq. ft to sq. m	0.09290
Sq. m to sq. ft	10.76
Feet to cm	30.48
Feet to m	0.3048
m to feet	3.281
m to yards	1.094

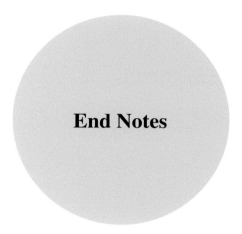

End Notes

Tools

- Tape Measure
- 2-ft level
- Line level
- Framing Square
- Combination Square
- Pencil
- Regular Wood Plane and Chamfer Plane
- Plumb Bob
- Staple Gun
- ½″ Staples
- Box Cutter
- Caulking Gun
- Long-Handled Shovel
- Rake
- Hoe
- Pick
- Concrete Tamper (you can make this)
- 2-gallon Plastic Bucket
- Wood or Metal Concrete Float
- Wheelbarrow
- Small Sledgehammer
- Mason's Twine
- Chalk and Chalkline

© Springer-Verlag New York 2016

J.S. Hicks, *Building a Roll-Off Roof or Dome Observatory*, The Patrick Moore
Practical Astronomy Series, DOI 10.1007/978-1-4939-3011-1

- Handsaw and Keyhole Saw
- Metal Hacksaw
- Steel saw
- Circular Saw
- Electric Drill and Bits
- Rechargeable Driver
- Brace and Bit with Bits (for longer bolt-holes)
- Claw Hammer
- Range of Screwdrivers
- Adjustable (Crescent) Wrench
- Various size Nail Sets
- Wood Chisel
- Short Pry Bar
- Miter box and Backsaw
- 6-ft Stepladder
- Work Gloves
- Safety Goggles
- Dust Mask
- Tin snips
- Wood clamp
- Ratchet wrench
- Spatula
- Portable Jig saw
- Post hole digger
- Electric plane
- Rivet gun + all-aluminum rivets

Nails

- 1″ Ardox
- 1-¼″ Ardox
- 1-½″ Ardox
- 2″ Ardox
- 2-½″ Ardox
- 3″ Ardox
- 3-½″ Ardox
- 1″ or 1-½″ Roofing Nails
- 2″ or 2-½″ Siding Nails

Screws

- Round-Head Wood
- Flat-Head Wood
- Hex 3″ × 3/8″ Lag Screws
- Robertson 3″ deck Screws

Bolts

- Anchor or "J-Bolt"
- Carriage Bolts
- Machine Bolts
- Flat head Stainless Steel $2'' \times \frac{1}{4}'' \times 20$ thread: 50 mm $\times$ 6 mm $\times$ 20 thread
- Round head Stainless Steel $1'' \times 8 \times 32$ thread: 25 mm $\times$ 8 $\times$ 32 thread

Fasteners/Hardware

- Joist Hangers
- Hurricane/Seismic Hangers
- Header Hangers
- Truss/Rafter Hangers
- Hinges
- Door Latch
- Pre-hung Door
- Door Knob and Deadbolt kit

Footings/Foundation

- ¾″ (19 mm) Crushed Gravel
- Concrete mix
- Polyethylene Sheet
- Preformed Sono-tubes (3050 mm diameter)
- Wood Shims (Shingles)
- 5/8″ (16 mm) T&G Plywood Floor Sheathing
- 5/8″ Steel Rebar (16 mm)
- $6'' \times 6''$ welded #8 mesh

Roofing/Siding

- Ply Wood Roof Sheathing (if deemed necessary by building code)
- Siding Panels/Boards (if not vinyl)
- Masonite sheathing to back up vinyl siding
- 15-lb Roofing Felt
- Vinyl Board and Batten Siding
- Ethafoam

Track and V-Groove Roller Hardware

- 4 sections each 15′ long—1-½″ × 1-½″ Steel track angle
- 4 sections each 15′ long—3-½″ × ¼″ Steel Plate
- 10—V-Groove Casters (Bestway Casters specified)
- 2 × 1000 lb Boat Winches with safety-dogs
- 125 ft 1/8″ Woven Stainless Steel Cable

- 2 Enclosed Sailboat Pulleys and Eyes
- Harken Fiddle/Becket Sailing Block (see Fig. 15.17 below)
- Eyelet Guides for Stainless Steel Cable
- 2 Pillow Blocks with 1″ Diameter RollerBearings or a Fairlead
- 6 ″×1″ diam. Steel Shaft (preferably Stainless Steel for Pillow Blocks)
- 1′ diam. Chain wheel and chain or a small Ship's wheel
- S/S Ball bearingbtransfer rollers
- S/S Piano hinge
- Stainless Boat snaps (4)
- Springs for arrestor
- 3″ diam.
- 3″ (75 mm) Pipe (foor flanges)
- Section of ½″ (12 mm) pipe
- 4 S/S Eyebolts (1‴ eye)
- Bungee chord
- 1–6″ (152 mm) turn buckle

Figure (Sailing Block): Harken/Becket sailing block allows rope to pass over it as the slot door proceeds to the top of the dome and allows the retraction of rope and door back over it as the slot door is closed [Coutresy of Harken, Wisconsin]

Painting Supplies

- Wide Brush
- Paint Tray
- Roller
- Spray Gun
- Wood Stain
- Two-part Epoxy cement
- Fiberglas resin
- Fiberglas cloth
- Tremclad metal spray
- Epoxy-resin acrylic base paint
- Marine Spar varnish
- Exterior Paint (If wood siding)
- Rags
- Silicone sealand caulking

Electrical

- 10 G underground insulated wire
- Electrical conduit and cement
- 1 Exterior Junction boxes
- 4 Interior Junction boxes

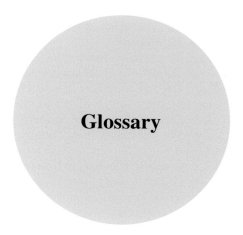

Glossary

Accessory building In zoning this refers to a building other than the dwelling on a lot. This would include a detached garage, a tool shed, or any outbuilding. The reason for this description in the by-law is simply to prevent occupation of the accessory building as a domicile (home), particularly before a home is built on the lot.

A-frame A type of building or structure that resembles an "A". Structurally, the frame of the building is in the form of an "A". In observatory design, it relates to a set of walls that form an "A" when closed.

Allen key screw A screw with a hexagonal socket that can be turned by socket wrenches and drivers—usually employed to grip on bearings, wheels, or knobs to prevent them from turning.

Ambient air (temp) Air that is at the same temperature as surrounding air. In the observatory, air that is the same temperature as the outside air. It is essential that the telescope optics be at the same temperature as the outside air.

Asphalt impregnated underlay A foundation fibre material that is saturated with asphalt that lies under wood frame construction to prevent insect, rodent and rot from moisture.

Airfoil A structure that allows 'lift' due to its design (like an aircraft wing).

Ball transfer rollers Usually a course of rollers raised slightly above a flat, level plane that can carry materials across the plane (usually will support massive weight).

Balloon framing A framing method where the wall studs extend below floor deck, attached directly to wood foundation members that are bolted to a concrete slab. Both the studs and first-floor joists rest on the foundation. The joists are nailed to these studs.

Base plane The lowest part of the ground surrounding the observatory.

Battens Narrow boards used to cover vertical joints in siding or wallboard.

Batter Boards A temporary board construction forming a right angle to which strings are attached for locating and squaring the corners of the observatory. It is usually offset from the foundation corners by at least 2 ft.

Bending mill (roller bending mill) A heavy steel mill comprised of rollers so arranged that they bend a shaft or angle of steel into a circular form under desired radius.

© Springer-Verlag New York 2016
J.S. Hicks, *Building a Roll-Off Roof or Dome Observatory*, The Patrick Moore
Practical Astronomy Series, DOI 10.1007/978-1-4939-3011-1

Bi-metal reaction The chemical erosion that occurs between specific metals on the metals activity list, particularly if they are touching.

Birds's-mouth (cut) A cutout in a rafter where it crosses the top plate of the wall, to provide a bearing surface for nailing. Due to its "triangular" notch, it requires a wider rafter than a $2'' \times 4''$. It can be eliminated with modern sheet steel "hangers".

Blocking Short sections of wood installed between framing studs, used to provide a nailing surface for boards or panel edges, or simply to reinforce the framing strength of walls.

Board-and-batten A type of board siding that consists of an initial layer of boards nailed with 2-½″ (8d) nails at 16″ on center along their centerline, leaving a gap between boards of ½″. The next layer of boards (the battens) are less than ¼ the width of the initial boards, nailed over the gaps in the initial boards with 3-¼″ (12d) nails. These nails should be driven between and not through the initial boards.

Boiler "punch-outs" Slabs of thick circular steel either cut by torch or "punched out" with a large press to make holes for fittings in a boiler. Usually from ½″ to 1″ thick they are perfect for pier caps if not curved. They can be either filed smooth around the edges or spun in a lathe to make a completely round platform. Found in scrap yards.

Bolt circle The exact pattern of bolts that are to pass through a pier cap (at the top of the pier) or pier flange (at the bottom of the pier). In this case, the reference is to creating a template in plywood to contain the bolts that anchor the pier into the cement footing underneath it.

Bottom chord The long $2'' \times 4''$ chord that runs from the caster plate-rafter connection across the roof bottom. Sometimes called the 'ceiling joist' if a ceiling is planned for.

Box tube A hollow steel beam that is used to contain the V-groove casters in an elaborate split-roof observatory model.

Bridging $1'' \times 2''$ lumber placed in an "X" pattern between floor joists to keep them from twisting or warping.

Building code The regulatory code or system of laws used by cities, towns and municipalities to control and regulate building procedures and practices.

Bungee cord An elastic strap used for tying loads (i.e. boats etc).

'Camera obscura' A term applied to a camera hidden within a walled structure, usually elevated on a roof. Often hidden in a cupola or hut to survey the town below without being seen.

Caster gap The open air gap created by the casters, including the track.

Caster plate The $2'' \times 6''$ rail that the casters are bolted to, that each rafter frame is also tied to with a framing anchor. The caster plate runs the full length of the roof under all the rafter frames, and supports the roof framework over the track.

Caster stops Wooden blocks bolted to the 'track plate' or uppermost $2'' \times 6''$ plate of the observatory wall. At the very ends of the track plate, they prevent to casters from moving further off the rack.

Caster ring In a Dome observatory, the casters plus plywood ring holding them that the dome sits upon.

Caster track The steel inverted angle that runs the length of the track plate, through the entire length of the observatory, upon which the casters ride. Nip-welded to a steel base plate which is screwed down to the track plate, it can be adjusted sideways slightly to allow for slight changes in the width of the track.

Centroid The false center of the dome which all the ribs point to (false because the slot opening occupies any real centroid).

Chain link A closable link which ties together the ends of a chain. It can also be used to link a shackle to a clip or snap.

Chain wheel An internally toothed wheel that carries an endless chain which loops around it. Its axle continues through one caster drive wheel. When the chain is pulled it rotates the chain wheel which in turn rotates the caster on the Dome track.

Channel-groove siding A type of wall siding that presents a grooved appearance created by an overly wide tongue which only partly fits into the adjacent board's channel creating a 'groove' at each joint. It should be nailed with 2-½″ siding nails (8d) at 16″ oc.

Chinking The stuffing of "chinks' or spaces in a log-cabin style structure with cement, mud, or plaster. In early days, cattle hair was added to the plaster to increase its adhesive consistency and hold it rigidly in place. 'Chinkless' log-cabin construction involves grooving the under side of every log in each tier so that it saddles the log beneath, making a close joint for its entire length. The log observatory in this book was built in this fashion.

Chord The bottom or top cleat running across the rafter frame joining the rafters at their tops or at their bottoms. In our case the bottom 2″×4″ cleat is the chord referred to, running from the track plate across to the opposite track plate. It is also the definition for any line dissecting a circle.

Clam-shell roof (design) A type of observatory with either twin roof sections that swing upward, or roof sections that nestle within each other. The former with twin roof sections, involves a counterweight arrangement to allow easy lifting. The latter, with roof sections that nestle within each other, are usually found on dome observatories where the roof sections nestle together to leave one-half of the dome open. Weather-sealing the roof sections at their junction when closed can be a problem.

CNC (machining) Robotic pre-programmed machining by computer.

Collar tie A horizontal cleat placed between rafters to reinforce their strength. Usually applied very near to the ridge board.

Concrete pier A full-length pier from footing to pier cap (attachment point for telescope mounts). A cement footing at least 4 ft in ground is poured first. Cement is then poured into a sono-tube form positioned vertically on top of the footing. The use of a full cement pier requires pre-calculation of the height of the telescope in its horizontal storage position in order that it not interfere with roof closing.

Counterweights Cement or iron weights used to counterbalance the weight of the clam-shell observatory roofs as they are swung open. These weights are made by trial and error usually by adding sand or water in a container until balance is achieved. At that point the container is weighed on a scale and comparative cement or iron weights are fabricated and bolted to the arms.

Crawl space The space under the floor joists in an elevated floor such as in the sono-tube footing type foundation.

Diagonal The line connecting opposite corners of the observatory foundation batter boards or the foundation itself. It is the hypotenuse of the right-angle triangle formed. The two diagonals (hypotenuse) should be exactly the same when measured, assuring squareness of the foundation.

Dome base ring The aluminum (or steel) angle that holds the arches and ribs at the base of the dome.

Door veneer In the case of a dome, the alternative to using aluminum panels, but employing a wooden veneer (usually birch).

Drip line The linear strip beneath a roll-off roof or curved dome where rainwater falls and carves out a narrow channel.

Drainage act A provincial, or state act that regulates the distribution of water over land, in our case from one property to another.

Eave The roof overhang which extends out from the observatory side walls.

Econcoat An insulated wall panel.

Electric plane A portable (sometimes rechargeable)—plane for planing wood.

Epoxy resin A two-part resin (resin and hardener) that forms a strong glue junction.

Epoxy resin—acrylic base An ideal coating for cement slab observatory basements.

Ethafoam A high density foam, easily shaped or cut for lining around the pier.

Exhaust fan A wall fan that can be built into the gable of the observatory to pull the hot air out of the observatory.

Eye bolt A heavy duty eye formed on a bolt shank used for tieing rope, wire, or clips to.

Fascia The finishing board applied over the end of the rafters at the eaves.

Filler Blocks Same as 'blocking', used alongside studs at wall ends to provide a nailing surface, and (also in the case of a corner) to provide extra strength and support.

Fiberglass resin A liquid resin normally reinforced with fiberglas cloth.

Fiberglas cloth A glass fiber cloth used to bind fiberglas resin to a substrate.

Flanges Circular (or square) steel platforms welded or bolted to both upper and lower ends of the steel pier. At the bottom, the flange allows bolting of the pier to the concrete pier footing, and at the top—to the telescope mount. Usually of steel plate turned on a lathe.

Footing The below-ground portion of a foundation, or a poured-concrete base upon which sono-tube concrete piers are set. A widened concrete footing sits beneath the metal pier extending below-ground.

Footprint The area covered by the building foundation, or first floor.

Foundation plan A plan showing only the basement section of a structure, or the area normally composed of block, stone or cement.

Framing square A steel square used by carpenters and contractors to calculate and lay out rafter lengths and right angle cut-offs.

Frost heave The shifting or upheaval of ground due to alternate freeze-thaw cycles, usually attributed to freezing water in the soil.

Frost line The maximum depth to which soil freezes in your climatic zone.

Gable The portion of roof overhanging the end walls of a gable roof structure.

Gable trim The metal finishing trim that is applied at the end of a gable (like a fascia board on the eaves portions). It is usually 'braked' (bent in a 'brake') to fit your particular gable-end.

Gable roof A pitched roof with two sides.

Galvalume An aluminum steel mixture used in dome panel manufacture.

Gantry The supporting framework for the track and roll-off roof beyond the end-wall of the observatory.

German Equatorial Mount The standard equatorial mount for telescopes required for star tracking and photography.

Grease nipple A grease point fitting on machinery, particularly with bearings.

Hand-held jig saw A portable jig saw guided by hand.

Harken Fiddle/Becket (Sailing Block) A special sailing block that can flip over carrying a rope or wire up to it and past it, and be retrieved back over it.

Hatch Used mostly in sailing jargon, the term refers to a removable, sliding, or swing-out small door. Our reference is to a small swing-open door in the roof of a shed or lower part of a slot door in a dome.

Header A horizontal framing member nailed across the ends of joists, usually to close the ends of joists flush with the outside observatory walls. Also framing over doors and windows.

Head jamb The top jamb in door framing, usually rabbeted (stepped channel) to accept the top of the door.

Helio arc welding A type of welding used to join/weld aluminum structures.

Hemisphere Half of a sphere.

Hip roof A pitched roof having four sides.

Hurricane/Seismic tie A metal bracket prefabricated to hold the rafters to a ridge board without toe-nailing the rafter to the ridge board. Although specifically made for Hurricane/Seismic-proof roofing construction, it offers a strong tie for a movable roof—offered by the Simpson Company.

Hypotenuse (diagonal) In our layout of the foundation, the diagonal between two opposite corners of either the batter boards or the edges of the footing(s).

Inverted angle Right-angle steel track that is welded leg-down with its back up to form the caster track (Roll-Off Roof Observatory).

J-Bolts Steel foundation bolts in the shape of a 'J' which are sunk into wet concrete and bolted to the sole plates of the wall sections.

Joists Horizontal framing members which support the floors or ceilings.

Joist-Hanger Prefabricated sheet metal hanger which holds a joist against a header or any other horizontal member. Nails are driven through tabs in the hanger which gives the unit exceptional strength—much preferable to butt-nailing or toe-nailing it.

Key Fob Transmitter An electronic key switch that operates machinery or apparatus from a short distance.

Kick-Rail A low railing meant to stop people advancement in a dangerous area.

King Post A vertical member that supports the ridge board from underneath, stretching from a bottom chord to the ridge board. It is often butted against a long chord running the length of the roof, that sits on the bottom chord that runs transversely *across* the roof.

Lag bolts (lag screw) Long screw-threaded bolts with a hex-head that are used for supporting heavier loads in wood-framing. More often used in square timber connections such as post and beam structures.

Laminar air flow Air that moves horizontally over a surface in a layer, such as over the surface of water.

Leg-in (angle) A length of metal angle curved to a specified inside radius.

Leg-out (angle) A length of metal angle curved to a specified outside radius.

Look-out A strut under the eaves of the roof running horizontally from the facia trim back to the wall where it is usually fastened to a ledger running parallel to the eaves. In our case, the look-out is a triangular section of $2'' \times 6''$ lumber cut in a triangle to terminate at its $6''$ width. Glued and strapped with hangers to the rafter it forms a member upon which the soffit (plywood or aluminum) can be nailed.

Louvres Narrow boards spaced and angled to stop rain etc., from entering, but allowing air to flow through, "usually in a fan duct".

Main use In zoning refers to a Dwelling Unit or Home versus an "Accessory Use" such as a shed or outbuilding.

Mortise A recess or cut-out in a board or member designed to receive the end or flank of another member.

Masonite A hardboard of pressed-wood composition used for underlay and sometimes as a finished wall board.

Metal pier The uppermost section of pier that the mount and telescope is bolted to. Usually large diameter steel or aluminum thick-walled pipe, with flanges welded to top and bottom for attaching telescope mount above, or cement Pier below.

Minor variance A granted variance to the Zoning Bylaw allowing a slight modification to the by-law (minor). Major deviations to the by-law require a Re-zoning or Zoning Amendment. The process requires an application to the "Committee of Adjustment" or the Town Council which is judged on its merits. An example would be a reduced side-yard width to accommodate our Roll-Off Observatory.. The appellant (owner of the parcel) would have to give good evidence of why he needed the variance. The application has merit if it does not affect the adjoining properties. It is not a question of size, area, length or width, but more of a question of impact on surrounding property.

Motorized drive A telescope driven by small motors in declination and equatorial planes.

Newtonian telescope A reflecting telescope with a mirror instead of a lens that focuses light to an eyepiece—refers specifically to a particular design invented by Isaac Newton.

Nip-Weld A slight weld which involves only a "spot" of welding applied at intervals to avoid warping the metals being joined.

Official plan The main legal "instrument" the Town, city, state, or municipality has to control development on its land base. It is a broad planning document which describes the intent of the Zoning By-law as it is applied to all land uses, and is implemented by the Zoning By-law. It usually avoids the deeper criteria involving measurements carried out specifically by the Zoning By-law. The Official Plan can be compared to the "Driver's Handbook" in the glove compartment of your car, whereas the Zoning By-law is the "Mechanic's Manual". An application for Official Plan Amendment may meet the size criteria of the Zoning By-law, but fail to meet the *intent* of the Official Plan which takes precedence over measurements.

Particle Board Particle Board or "Wafer-board" is a compressed wood fiber product manufactured in sheets. It is used largely for underlay on floors, roofs and walls. Since it is affected by water soaking into it, I prefer it only used on vertical walls where a water soaking will simply run off or evaporate quickly. It is not a good nailing surface, but rather only an intermediate layer usually sandwiched between outside boarding and the studs of a wall. I do not advise its use on roofs or floors. Tongue and Groove ply is safest on floors and roofs.

Perforated tile A plastic ribbed drain tile usually 4″–6″ diameter with a series of slits in its underside which allow water to enter it and thus drain it off by gravity.

Pergola An arbor usually covered with vines or flowers trailing over a trellis, or often just a open joist structure which lets sunlight through.

Piano hinge A long narrow hinge normally used in piano construction to hold the lid on the top of the piano.

Pier In this book, pier refers to a telescope foundation in the form of a concrete form or a steel tube supported by a concrete form.

Pillow block Refers to a large bearing or set of bearings through which a steel shaft runs to carry cable over it and beyond, acting like a large, wide pulley.

Pipe floor flanges Flat bottom flanges that thread into plumbing pipe used for railing staunchons. They can be screwed down to a floor and used as railing posts.

Platform framing A framing method that involves building a plywood floor on which the walls are erected (see balloon framing).

Polar alignment The art of locating the declination axis of the telescope along the polar axis (pointing to the star polaris).

Portable driver Referring to a motorized chargeable and portable screw-driver.

Pulsar Observatory A manufactured observatory in the UK distributed in North America.

Plumb Vertical, or to make vertical.

Plumb cut A vertical cut made on the top end of a rafter to meet the ridge board at the top of the roof.

Polyethylene A plastic polymer of ethylene used in containers, sheating & packaging.

Post hole digger A clam shovel used by hand top dig holes, or a motorized auger.

Purlins Narrow boards running across the roof rafters for support of the roof sheathing or steel sheet roofing. Either mortised into the rafter if it is wide enough or framed into the rafter with metal anchors.

PZT A platform (shelf) used in the Sky Shed POD observatory allowing the dome to be pushed off its track slightly, permitting a view past the Zenith.

Rafter Framing members used to support the roof.

Rafter Hanger A pre-formed sheet metal support used in lieu of butt-nailing or toe-nailing rafters to either ridge pole or top plate.

Ratchet wrench A type of wrench that has an adjustable one-way torque.

Rebar A ridged iron bar usually ½″–5/8″ diameter used to reinforce concrete footings. Usually laid within a cement pour about mid-level of the pour.

Reinforced concrete Concrete containing either wire (steel) mesh or rebar.

Rebar cage A basket-like column of rebar tied together with heavy wire which runs the length of a cement pier or foundation.

Receptacle An electrical outlet box usually containing twin plug outlets. Sometimes called a 'duplex'.

Ridge Board (Ridge Beam) The upper-most portion of a roof structure. A beam running the whole length of the roof at the peak with roof rafters nailed into it.

Ridge cap A pre-formed metal cap that runs along a steel roof peak. It is constructed to allow air to flow through it ventilating the steel sheets.

Rivet (pop-rivet) A type of compressable bolt, with a headless end, widened by a tool that squeezes and widens it to make it immovable.

Rivet gun The hand-operated tool that compresses a rivet in place.

Rise/run A formula representing the pitch of a roof expressed as a ratio. For instance a rise of 4 in. for every 12 in. run (horizontal) represents a 4-in-12 pitch.

Robertson screws Screws with a square recess to fit a square-tipped screwdriver.

Roller fairlead A type of bearing that is elongated and lies flat on a surface to guide rope or wire over a barrier. Often used where a change in angle is required (i.e. over a barrier).

Rolling ladder A steel ladder in the form of steps on a platform with casters underneath which can be rolled around. It can be fixed in place with a brake on the casters. Enables you to sit on the upper stair (which is a platform) at a convenient height for viewing through a telescope.

Roll-off roof A roof which rolls or slides off a structure revealing an open room.

Roof felt An asphalt-saturated material used under shingles or steel roofing or for water-proofing. It serves to prevent condensation under a steel roof undergoing rapid cooling.

Roof framing member A complete rafter assembly section—rafters, ridge beam, collar ties or chord tied together as a whole.

Roof pitch See rise/run.

Roof sheathing Covering over the rafters in the form of plywood or particle board etc. Sometimes referred to as the 'roof deck'.

Saddles (iron saddles) Heavy guage steel brackets pre-formed to fit various size timber posts. Variations in design include flat base with holes to be bolted into set concrete, or with a welded post which is to be imbedded into a wet concrete form.

Schmidt-Cassegrain (telescope) An enclosed mirror reflector with a large corrector lens in front holding a secondary mirror and a primary main mirror. This arrangement shortens the tube length. The primary mirror has a hole in it passing the focus to a point behind it.

Screed Board A wide board that is pulled across a wet cement floor to level it.

Shim strips Thin aluminum strims placed under the track to level it.

Ships wheel A spoked wheel in wood, found usually on yaghts and vessels.

Shutter The moveable observing slot on dome observatories.

'S'—Hooks S/S hooks that are used to attach chain together or to hang items from.

Sidereal (motion) The telescope tracking motion that follows the rotation of the stars.

Side jam The vertical member or stop at the sides of a door frame.

Sideyard The area of land between a building and the lot line on the same side.

Site plan A plan describing a survey of the site, and all its features (often including buildings).

Silicone sealant A silicone cement in a tube that is squeezed out along seams.

Silo Dome A dome prefabricated in panels by farm silo manufacturers.

Skirt A portion of roof overhang that extends vertically down the wall. Usually no more than a foot long, it is advantageous for covering and weatherproofing the caster gap, normally an area open to the air.

SkyShed MAX An observatory marketed by SkyShed Observatories.

SkyShed POD A new observatory replicating a dome, by Sky Shed Observatories.

Slab In our text, it refers to a full concrete floor in the observatory section or a continuous footing under the gantry section.

Slewing The motion of the telescope driven to follow the stars in their motion, or the motion of the telescope moved and searching across the sky.

Slot Door The moveable section that slides up and over the dome.

Slot Door brace One of three braces that hold the two halves of dome together (the Slot Door Stop is one of them).

Slot Door guides The leg-up aluminum angles that guide the door up and over the dome.

Slot Door stop The horizontal cross member at the dome top that stops the slot door from sliding off the frontside of the dome (the slot door top end comes to rest upon it).

Soffits The portions of the rolling roof overhanging on the sides of the observatory that are sheathed underneath to enclose the rafter framework.

Soil profile Layers of soil including the soil horizon usually visible in a dug hole. For instance a clay soil profile would indicate a high concentration of clay and clay-based ingredients.

Solar panel charger A type of solar panel charger that is portable for small consumtion.

Sole plate The lower 'plate' or member on a wall section. The joists are usually butt-nailed through this plate when the wall framework is being assembled on the ground. When the wall framework is hoisted up onto the plywood floor deck the sole plate is nailed to the floor underneath.

Sono-tube A rolled fiber tube used for forms in footings. It is spiral-wound and usually waxed so that if the soil heaves from frost, it will slide by the tube without lifting the footing. (Hence the necessity of placing the bottom of the sono-tube well below the frost-line where it could be lifted by vertical forces on the bottom.)

Span The distance between vertical supports for horizontal framing members, such as joists or beams. For rafters it is the distance between two opposite walls that support them, or sometimes the span of a building.

Spoked jig A pre-welded jig that hols the caster track in a perfect circle, later removed.

Spring-loaded arrester A double spring set in a channel on the backside of the dome that "arrests' the fast motion of the slot door, preventing damage to the dome, slot door or operator.

Squatting Residing on land (either in a home or structure) without permission and particularly without legal ownership or a deed.

Standing (box) rib A type of rib manufactured by silo dome manufacturers for domes.

Steel saw A circular blade used in a Skill Saw for cutting metal.

Studs The vertical members of a building forming the framework for the walls.

Tail cut The bottom vertical cut at the lower end of the rafters where a facia board will is usually nailed.

Tapping The action of threading a hole for a bolt accomplished with a 'tap' (threading tool).

Tarn A small mountain Lake formed by meltwater.

Telephone dialer An automated dialer set off by an alarm that dials police or a security guard.

Template In our text this refers to a temporary rafter member, or test rafter, temporarily assembled to verify its proper dimensions to fit the 'span'.

Thermal equilibrium When a telescope has reached the same temperature as the outside surrounding air.

Thermal window If installed between 'warm room' and observing room this refers to a window that is well-insulated to keep any heat supporting the warm room from leaking into the observatory.

Threshold The portion of a door or 'sill' that lies at its base, usually replacing the sole plate in that gap.

Trickle charge A low voltage, slow charger that charges a battery.

Top plate The upper-most part of a wall section, usually a double plate. In the Roll-Off observatory the double top plate carries the track. The top plate of the two is referred to as the 'track plate' in our text and diagrams.

Torque A rotational force (usually on a bearing or wheel).

Track gap The air gap created between the track plate and the caster plate above it. Essentially the total gap formed by the casters plus track.

Track Joist A joist mortised into the top of the wall studs that runs the full length of the observatory (walls plus gantry). It aids in supporting the weight of the roof both on the observatory section and over the gantry section.

Track plate The upper top plate that carries the track rail running the full length of the observatory structure (from walls through gantry). Part of it sits over the track joist underneath.

Trellis A wooden garden lattice structure. In our case it is installed as a partially-open ceiling between the gantry joists to clear the rolling roof.

Tremclad A rust proof protective coating applied to metal.

Truss A reinforced rafter member, pre-manufactured to fit your span. It is held together with a nailing plate to support a specific weight. You could have such a member made up by your local lumber company avoiding the trouble to make your own rafter members.

Turbulence The effect of rising hot air. Detrimental to viewing with optical aid such as with a telescope.

Turn-buckle A spindle-shaped device with threads and eyes (or a hook) on both ends, that is used for tightening rope or wire (commonly found on sailboats and clotheslines).

Two-part epoxy An epoxy cement in two tubes containing resin and hardener.

Variance See 'minor variance'.

Ventilating fan A fan inserted in the gable of a roll-off roof observatory, or wall of a dome observatory. Used to usher hot air out of the structure.

V-Groove casters A type of cast steel roller that is designed to roll on a 90° knife-edge track such as an angle iron mounted with its back up.

Warm room A separate room partitioned off from the regular portion of the observatory for warm, relaxed viewing by means of a digital camera and lead to a computer screen. Most often also a 'dark room' to make the monitor screen visible.

White room A 'clean' room deriving its name and appearance from clean, white walls in a dust-free, uncontaminated environment. A room with extraordinary cleanliness.

Worm-gear Referring to a worm-gear drive system on a telescope mount.

Wood clamp A large clamp normally used to keep glued wooden parts together.

Zenith The point in the sky directly above an observer. The point directly above the center of the floor of the observatory.

Zoning The legal 'instrument' through which a state, town, or municipality controls its development—its' purpose is to create 'orderly' development through a series of numerical standards for yards, areas, lot sizes etc.

Zoning amendment An alteration or change to the Zoning By-law to allow something that is not normally permitted in that zone. It requires an application and review by town or city council who will judge it as appropriate or non-conforming. You want to avoid having to go through this costly, and risky business, if at all possible.

Zoning by-law A formal document that outlines all the types of zoning categories within a state, town, or municipality. It also contains a list of zoning amendments or exceptions to the zoning by-law.

Index

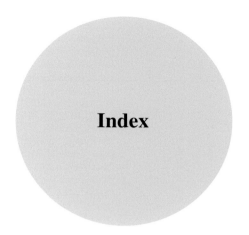

© Springer-Verlag New York 2016
J.S. Hicks, *Building a Roll-Off Roof or Dome Observatory*, The Patrick Moore
Practical Astronomy Series, DOI 10.1007/978-1-4939-3011-1